Advanced Graphene and Graphene Oxide Materials

Advanced Graphene and Graphene Oxide Materials

Editors

Victoria Samanidou
Eleni Deliyanni

Basel • Beijing • Wuhan • Barcelona • Belgrade • Novi Sad • Cluj • Manchester

Editors
Victoria Samanidou
Chemistry- Laboratory of
Analytical Chemistry
University of Thessaloniki
Thessaloniki
Greece

Eleni Deliyanni
Chemistry
University of Thessaloniki
Thessaloniki
Greece

Editorial Office
MDPI
St. Alban-Anlage 66
4052 Basel, Switzerland

This is a reprint of articles from the Special Issue published online in the open access journal *Materials* (ISSN 1996-1944) (available at: www.mdpi.com/journal/materials/special_issues/ Graphene_Graphene_Oxide_Materials).

For citation purposes, cite each article independently as indicated on the article page online and as indicated below:

Lastname, A.A.; Lastname, B.B. Article Title. *Journal Name* **Year**, *Volume Number*, Page Range.

ISBN 978-3-7258-0008-7 (Hbk)
ISBN 978-3-7258-0007-0 (PDF)
doi.org/10.3390/books978-3-7258-0007-0

© 2024 by the authors. Articles in this book are Open Access and distributed under the Creative Commons Attribution (CC BY) license. The book as a whole is distributed by MDPI under the terms and conditions of the Creative Commons Attribution-NonCommercial-NoDerivs (CC BY-NC-ND) license.

Contents

About the Editors vii

Preface ix

Mallikarjun Madagalam, Mattia Bartoli and Alberto Tagliaferro
A Short Overview on Graphene and Graphene-Related Materials for Electrochemical Gas Sensing
Reprinted from: *Materials* **2024**, *17*, 303, doi:10.3390/ma17020303 1

Manoel L. Silva-Neto, Renato Barbosa-Silva, Georges Boudebs and Cid B. de Araújo
Second-Order Nonlinearity of Graphene Quantum Dots Measured by Hyper-Rayleigh Scattering
Reprinted from: *Materials* **2023**, *16*, 7376, doi:10.3390/ma16237376 21

Tikaram Neupane, Uma Poudyal, Bagher Tabibi, Wan-Joong Kim and Felix Jaetae Seo
Cubic Nonlinearity of Graphene-Oxide Monolayer
Reprinted from: *Materials* **2023**, *16*, 6664, doi:10.3390/ma16206664 34

Francesco Inchingolo, Angelo Michele Inchingolo, Giulia Latini, Giulia Palmieri, Chiara Di Pede, Irma Trilli, et al.
Application of Graphene Oxide in Oral Surgery: A Systematic Review
Reprinted from: *Materials* **2023**, *16*, 6293, doi:10.3390/ma16186293 44

Erik Biehler, Qui Quach and Tarek M. Abdel-Fattah
Gold Nanoparticles AuNP Decorated on Fused Graphene-like Materials for Application in a Hydrogen Generation
Reprinted from: *Materials* **2023**, *16*, 4779, doi:10.3390/ma16134779 63

Xiaojiang Hong, Jin Chai Lee, Jing Lin Ng, Zeety Md Yusof, Qian He and Qiansha Li
Effect of Graphene Oxide on the Mechanical Properties and Durability of High-Strength Lightweight Concrete Containing Shale Ceramsite
Reprinted from: *Materials* **2023**, *16*, 2756, doi:10.3390/ma16072756 76

Song Mi Kim, Woo Rim Park, Jun Seok Park, Sang Min Song and Oh Heon Kwon
Effect of Defects in Graphene/Cu Composites on the Density of States
Reprinted from: *Materials* **2023**, *16*, 962, doi:10.3390/ma16030962 94

Rabia Ikram, Badrul Mohamed Jan, Waqas Ahmad, Akhmal Sidek, Mudasar Khan and George Kenanakis
Rheological Investigation of Welding Waste-Derived Graphene Oxide in Water-Based Drilling Fluids
Reprinted from: *Materials* **2022**, *15*, 8266, doi:10.3390/ma15228266 106

Dipendra Dahal, Godfrey Gumbs, Andrii Iurov and Chin-Sen Ting
Plasmon Damping Rates in Coulomb-Coupled 2D Layers in a Heterostructure
Reprinted from: *Materials* **2022**, *15*, 7964, doi:10.3390/ma15227964 122

Aamir Razaq, Faiza Bibi, Xiaoxiao Zheng, Raffaello Papadakis, Syed Hassan Mujtaba Jafri and Hu Li
Review on Graphene-, Graphene Oxide-, Reduced Graphene Oxide-Based Flexible Composites: From Fabrication to Applications
Reprinted from: *Materials* **2022**, *15*, 1012, doi:10.3390/ma15031012 135

Vladimir P. Vasiliev, Roman A. Manzhos, Valeriy K. Kochergin, Alexander G. Krivenko, Eugene N. Kabachkov, Alexander V. Kulikov, et al.
A Facile Synthesis of Noble-Metal-Free Catalyst Based on Nitrogen Doped Graphene Oxide for Oxygen Reduction Reaction
Reprinted from: *Materials* **2022**, *15*, 821, doi:10.3390/ma15030821 . . . . **152**

About the Editors

Victoria Samanidou

Dr Victoria Samanidou FRSC is a Full Professor and Director of the Laboratory of Analytical Chemistry in the School of Chemistry of Aristotle University of Thessaloniki, Greece. Since 2022, she has been also Vice-President of the School of Chemistry. Her research interests focus on the development of sample preparation methods using sorptive extraction prior to chromatographic analysis in accordance to green chemistry demands.She has co-authored 219 original research articles in peer-reviewed journals and 67 reviews, 88 editorials/in view/opinions/commentaries and 60 chapters in scientific books (h-index 46, Scopus Author ID 7003896015, 7126 citations). She is an editorial board member of more than 31 scientific journals and guest editor in more than 32 Special Issues. She has peer reviewed more than 819 manuscripts for 171 scientific journals. In 2016, she was included in top 50 power list of women in Analytical Science, as proposed by Texere Publishers. In 2021, she was included in the "The Analytical Scientist" 2021 Power List of top 100 influential people in analytical science. In 2023, she was included in the Power List 2023, which is created in the spirit of the journal's recent 10-year anniversary of The Analytical Scientist Magazine and includes nominees who were assessed for their excellence and impact over the past decade across four categories: Innovators and Trailblazers, Leaders and Advocates, Connectors and Interdisciplinarians, and Mentors and Educators. The list is created of 100 scientists (25 per category). In this list, she was listed in the category of Mentors and Educators. In the last years, she has been included in the list of World Top 2% Scientists, for single years and career as well, published in PLOS Biology and based on citations from SCOPUS. She is also the Leader of Working Group 1 Science and Fundamentals of EuChemS-DAC Sample Preparation Study Group and Network (2021).

Eleni Deliyanni

Dr. Eleni A. Deliyanni was a Full Professor in the Laboratory of Chemical and Environmental Technology in the School of Chemistry of the Aristotle University of Thessaloniki (AUTh), Greece (retired in 2022). From 20/12/85 to 30/3/86, she was part of the Technical University of München, Germany, for additional work on her PhD Thesis, as well as research work on Scanning Electron Microscopy. From 1/12/07 to 30/4/08, she worked at the Institute of Analytical Chemistry at The City College of New York of CUNY, Department of Chemistry and Biochemistry in New York, while from 1/1/10 to 30/5/10, she returned as a visiting professor at the same lab in New York. From 1/9/14 to 31/10/14, she visited the Environcentrum Company in Kosice, Slovakia within the frame of the collaborative program, "water and soil clean-up from mixed contaminants". Dr. Deliyanni has co-authored more than 100 original research and review articles in peer-reviewed journals with an H-index of 42, and her work has been cited more than 5200 times. She is a member of editorial boards of scientific journals like Journal of Colloid and Interface Science, Elsevier, while she has reviewed more than 550 manuscripts for more than 100 scientific journals. She has also been a Guest Editor for several Special Issues in scientific journals. Her research interests include (nano)materials chemistry, materials characterization, environmental chemistry, nanotechnology, water and waste-water treatment, the desulfurization of gaseous streams or (bio)fuels, adsorption, catalysis, advanced oxidation processes (AOPs), activated (nano)porous carbons and graphene derivatives.

Preface

Graphene and graphene oxide are widely applied as successful sorbent materials for various compounds obtained from biosamples and surface water samples. Therefore, they are suitable for future use in numerous biomedical and environmental applications.

Moreover, their functionalization with magnetic nanoparticles can lead to magnetic sorbents, thus allowing convenient sample treatment via magnetic separation.

To date, a plethora of graphene and graphene oxide materials have been synthesized and successfully employed for solid-phase extraction of organic compounds from environmental and biological samples. The unique properties of these materials enrich the analytical toolbox available for the analysis of various organic compounds in various matrices and make them precise and valuable means for handling analytical and environmental issues.

This Special Issue was supported by the Sample Preparation Study Group and Network, supported by the Division of Analytical Chemistry of the European Chemical Society.

Eleven manuscripts, namely seven research articles, three reviews and one communication are included in this SI.

The Guest Editors wish to thank all authors for their fine contribution and all reviewers who improved the quality of submissions with their valuable criticism.

Victoria Samanidou and Eleni Deliyanni
Editors

Review

A Short Overview on Graphene and Graphene-Related Materials for Electrochemical Gas Sensing

Mallikarjun Madagalam [1,2], Mattia Bartoli [2,3,*] and Alberto Tagliaferro [1,4,*]

1. Department of Applied Science and Technology, Politecnico di Torino, Duca degli Abruzzi 24, 10129 Turin, Italy; mallikarjun.madagalam@polito.it
2. National Interuniversity Consortium of Materials Science and Technology (INSTM), Via Giuseppe Giusti, 9, 50121 Florence, Italy
3. Center for Sustainable Future Technologies (CSFT), Istituto Italiano di Tecnologia (IIT), Via Livorno 60, 10144 Turin, Italy
4. Faculty of Science, OntarioTech University, Simcoe Street North, Oshawa, ON L1G 0C5, Canada
* Correspondence: mattia.bartoli@iit.it (M.B.); alberto.tagliaferro@polito.it (A.T.); Tel.: +39-011-090-3400 (M.B.); +39-011-090-7347 (A.T.)

Abstract: The development of new and high-performing electrode materials for sensing applications is one of the most intriguing and challenging research fields. There are several ways to approach this matter, but the use of nanostructured surfaces is among the most promising and highest performing. Graphene and graphene-related materials have contributed to spreading nanoscience across several fields in which the combination of morphological and electronic properties exploit their outstanding electrochemical properties. In this review, we discuss the use of graphene and graphene-like materials to produce gas sensors, highlighting the most relevant and new advancements in the field, with a particular focus on the interaction between the gases and the materials.

Keywords: graphene derivatives; electrochemical sensing; graphene tailoring

Citation: Madagalam, M.; Bartoli, M.; Tagliaferro, A. A Short Overview on Graphene and Graphene-Related Materials for Electrochemical Gas Sensing. *Materials* **2024**, *17*, 303. https://doi.org/10.3390/ma17020303

Academic Editor: Irina V. Antonova

Received: 6 December 2023
Revised: 5 January 2024
Accepted: 5 January 2024
Published: 7 January 2024

Copyright: © 2024 by the authors. Licensee MDPI, Basel, Switzerland. This article is an open access article distributed under the terms and conditions of the Creative Commons Attribution (CC BY) license (https://creativecommons.org/licenses/by/4.0/).

1. Introduction

The production of highly sensitive materials for electrochemical sensing is a matter of great relevance for analytic science. Actually, research is focusing on finding the best trade-off between the performance and the toughness of electrode materials [1]. In this field, graphene and graphene-related materials (GRMs) can play a game-changing role.

The outstanding electrical properties of graphene combined with its superior mechanical and optical properties have attracted great interest in electrochemical sensing applications due to the achievable sensitivity, rapid response times, and versatility in detecting a wide range of analytes never reached before [2]. The integration of graphene into electrochemical sensors has led to significant performance improvements in several fields of application, including environmental science [3], medical diagnostics [4] and quality control [5]. Graphene's astonishing performance is due to the improvement of electron transfer at the electrode interface by a combination of electronic and chemical features [6]. Furthermore, graphene and GRMs' high mechanical strength and flexibility further contribute to their utility in electrochemical sensing, ensuring the stability and longevity of sensors even under challenging conditions [7].

The tunability of graphene and GRMs allows their use across several types of sensors, including amperometric, potentiometric, and impedimetric ones. In amperometric sensing, the current generated by the electrochemical reaction at the electrode surface is measured and correlated with the concentration of the analyte. Graphene and GRMs contribute to sensitivity increments due to the combination of electrical conductivity and large surface area [8,9]. Potentiometric sensors measure the potential difference between a reference electrode and a working electrode, and the incorporation of graphene and GRMs enhance the stability and selectivity of the sensor, allowing them to be used in pH sensing [10] or ion detection [11]. Impedimetric sensors exploit changes in the impedance of the electrode interface upon interaction with the target analyte, and graphene's conductivity and charge transport properties increase both the response rate and the sensitivity of impedimetric sensors [12]. Among all the possible applications, gas detection represents a critical application of GRMs in electrochemical sensing due to the great deal of attention that monitoring air quality [13] and ensuring workplace safety [14] have gained. GRM-based sensors are of particular interest due to their ability to selectively interact with specific gases that are able to induce changes in the electric signals detected [15]. This ability, together with the other outstanding properties of GRMs, is of paramount relevance for a new generation of highly sensitive tough materials for multiple gas sensing.

In this work, we report the most relevant achievements in the GRM-based electrodes field, focusing on pristine graphene, graphene oxide (GO) and reduced graphene oxide (rGO) and their tailored derivatives. We summarize the key electronic properties of graphene and GRMs and diffusely discuss their applications in gas sensing applications, focusing on CO_2, CO, H_2, NH_3, NO_x, H_2S and SO_2. We provide a concise and easy-to-be-exploited overview aimed to represent a reference point for researchers interested in approaching electrochemical sensing using neat and tailored graphene and GRMs in gas sensing.

2. Graphene and GRM Electrical Properties

In agreement with the International Union for Pure and Applied Chemistry Golden Book, graphene is defined as "a single carbon layer of the graphite structure, describing its nature by analogy to a polycyclic aromatic hydrocarbon of quasi infinite size" [16]. A pristine graphene layer is composed of a planar arrangement of carbon atoms bonded through three σ bonds with the p orbitals perpendicular to the sp^2 plane, allowing a full delocalization of the π bonds [17–19]. This is the reason behind graphene's exceptional electrical properties, particularly its in-plane electron mobility. At room temperature, the electron mobility in graphene can reach up to 15,000 cm^2 V^{-1}s^{-1} [20] due to a peculiar band organization formed by two conical points in the electronic band diagram known as Dirac points, as shown in Figure 1.

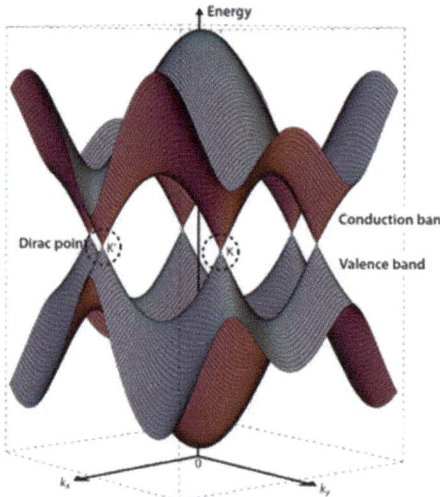

Figure 1. Three-dimensional schematic diagram of band structure near the Fermi level of graphene with Dirac points K and K′ highlighted. Reproduced, adapted and reprinted with permission from Lavagna et al. [21].

The linear dispersion around Dirac points contributes to the massless nature of charge carriers in graphene, allowing them to travel at incredibly high speeds without any significant scattering due to the eventual topological disorder due to the temperature [22]. Furthermore, graphene is characterized by a relevant Hall effect, with the plateaus occurring at half integers of $4\,e^2/h$ rather than $4\,e^2/h$. Near-ideal graphene sheets show a pronounced Hall effect, while multilayer samples show a much weaker gate dependence due to the electric field screening promoted by the other layers [23]. Interestingly, the use of a high magnetic field combined with cryogenic temperatures induces a quantum Hall effect for both holes and electrons [24,25]. The superb electrical properties of graphene are, however, counterbalanced by the absence of a band gap. A great effort has been devoted to creating graphene-like materials with a proper band gap [26], and solutions such as graphene nanoribbons have been developed [27]. Graphene nanoribbons can be shrunk by modifying the charge carrier momentum in the transverse direction, resulting in a band gap opening based on the ribbon width [28]. Alternatively, graphene can be doped with nanostructures and heteroatoms [29].

The most popular and useful procedure to dope graphene is oxidation with the formation of GO. GO is rich in oxygen functionalities such as epoxide and hydroxyl groups on its basal lattice, while carbonyl and carboxylic residues are more abundant on the edges, as described by the Lerf-Klinowski model [30]. The electronic properties of GO are strictly related to the degree of oxidation, as reported by Krishnamoorthy et al. [31]. The increment of GO oxidation induced a reduction of electron mobility [32] as a consequence of the introduction of more defects in the lattice structure of graphene. These defects act as scattering centers for charge carriers, hindering the smooth movement of electrons through the material. Nevertheless, the relationship between oxidation degree and electron mobility is a complex mix of factors, such as the oxygen functional groups and their distribution on the graphene lattice. Interestingly, GO shows a band gap due to the presence of the same defect that reduces the charge carrier's mobility. The oxygen functionalities alter the electronic configuration of the graphene plane, disrupting the π-conjugated system and forming localized states within the energy band structure of GO, giving rise to a band gap [33]. The presence of a band gap in GO promotes a semiconducting behavior contrary to pristine graphene that shows metallic conductivity [34]. This semiconducting behavior makes graphene oxide well-suited for applications in electronic devices where a controllable on/off state is essential [35]. An interesting compromise between the conductivity of pristine graphene and the properties of GO is the reduced form of GO, named rGO. rGO is produced using harsh reductive processes [36] for decreasing the oxygen residues of GO and trying to find a balance between pristine graphene and GO with a carbon/oxygen ratio ranging from 0.4 up to 13 wt% [37]. The electrical properties of rGO are far higher than those of GO but considerably inferior to graphene, while dispersibility showed an opposite trend [38].

3. Graphene and Graphene-Related Materials' Electrochemical Sensing Performance

Graphene and GRMs show several key features that allowed the spread of their use in electrochemical sensing applications. Firstly, GRMs are highly sensitive to the surrounding chemical and physical environment [39–41]. This is of particular interest considering the interaction with gas molecules that are adsorbed on GRMs' surface [42] that are able to alter the electronic conductivity [43]. The conductivity alteration induced by adsorbed gas prevents the correct interaction between the dangling π-orbitals and neighboring atom orbitals, altering the conduction bands and reducing the charge carrier's mobility. This phenomenon can be used to quantify the number of adsorbed molecules through simple electrochemical measurements in which GRMs represent the working electrode. Furthermore, the interactions between gaseous molecules and GRMs can be easily tuned by tuning the graphene functionalization, increasing both the electrochemical performance and the selectivity of the system. These features, together with a fast response and recovery of the electrodes, have boosted the use of GRMs as electrochemical gas sensing platforms, as summarized in Table 1.

Table 1. Overview of the key features of graphene and GRMs in electrochemical gas sensing.

Gas	Material	Highlights	References
CO_2	Chemical vapor-deposited single-layer graphene	■ Poor responsivity. ■ Negligible sensitivity in presence of water. ■ Linear range up 2000 ppm.	[44]
	Exfoliated graphite nanoplatelet	■ Good sensitivity in presence of water. ■ Linear range 10–200 ppm.	[45]
	Double-layer graphene	■ No cross-sensitivity with H_2O up to 3% relative humidity.	[46]
	GO	■ Cheap. ■ Robust. ■ Linear range 400–4000 ppm.	[47]
	rGO	■ Response of up to 71% in N_2. ■ Response of up to 15% in air.	[48]
	Inorganics (zinc, titanium, molybdenum)-tailored graphene	■ Operating from 10 to 60 °C. ■ Operating up to 97% of relative humidity. ■ Linear range 300–1100 ppm.	[49]
CO	Palladium-tailored rGO	■ Operating at 150 °C. ■ Operating up to 71% of relative humidity. ■ Linear range 200–1100 ppm.	[50]
	Palladium and tin oxide-tailored rGO	■ Slow response rate of 70 s. ■ Operating up to 85% of relative humidity due to the formation of surface channels. ■ Linear range up to 400 ppm.	[51]
	Zinc oxide-tailored rGO	■ Fast response up to 9 s. ■ Response of up to 82%. ■ Recovery time of 14 s. ■ Linear range 1–1000 ppm.	[52]
	Manganese oxide-tailored rGO	■ Fast response up to 3 s. ■ Response up to 70 s. ■ Linear range 1–1000 ppm.	[53]
	Tin oxide-tailored rGO	■ Good selectivity over ammonia, hydrogen, and water at 25 °C.	[54]
	Nickel manganate rod-tailored rGO	■ Ultra-low detection limit (0.6–1 ppm). ■ Linear range 20–250 ppm.	[55]
	Copper oxide-tailored rGO	■ Ultra-low detection limit (0.3 ppm). ■ Slow response rate up to 76 s. ■ Slow recovery up to 247 s.	[56]
	Cobalt and iron oxide-tailored rGO	■ Fast response up to 0.5 s. ■ Linear range 10–40,000 ppm.	[57]
	rGO	■ Response up to 71% with 30 ppm of CO. ■ Sensitivity 10 ppm. ■ Recovery time of up to 30 s.	[58]
	Poly(3,4-ethylenedioxythiophene)-tailored GO	■ Recovery of up to 42 s. ■ Linear range 20–270 ppm.	[59,60]
	Poly(N-methyl pyrrole)-tailored rGO	■ Recovery of up to 36 s. ■ Detection limit of 1 ppm. ■ Linear range 10–275 ppm.	[61]
	Metal organic framework-tailored rGO	■ Fast response up to 30 s. ■ Fast recovery up to 70 s. ■ Great durability in CO atmosphere for over 30 days. ■ Sensitivity of 25 ppm.	[62]

Table 1. Cont.

Gas	Material	Highlights	References
H_2	Palladium nanoparticles onto single-layer graphene	■ Response of 33% in 1000 ppm of at H_2 25 °C. ■ Sensitivity of 20 ppm.	[63]
	Palladium nanoparticles onto 3D-GRMs	■ Response of 41.9% under 3% H_2 at 25 °C. ■ Easy to produce.	[64]
	Platinum nanoparticle-decorated rGO	■ Response of 8% under 0.5% H_2 at 50 °C. ■ Fast recovery up to 104 s.	[65]
	Platinum nanoparticles onto 3D-GRMs	■ High sensitivity. ■ Fast response up to 9 s. ■ Fast recovery up to 10 s. ■ Good linearity in the range from 1 to 100 ppm.	[66]
	Tin oxide onto platinum nanoparticle-decorated rGO	■ Enhanced response compared with palladium-decorated rGO. ■ Enhanced sensitivity compared with palladium-decorated rGO.	[67]
	Tungsten-decorated GO	■ Response of 50 mV in presence of H_2 (100 ppm). ■ Use in air atmosphere. ■ Detection limit of 11 ppm.	[68]
	Zinc oxide-decorated GO	■ Fast response up to 114 s. ■ Short recovery time up to 30 s. ■ Detection limit of 4 ppm.	[69]
H_2O	Vertically aligned graphene arrays	■ Sensitivity related to relative humidity. ■ Improved performance for relative humidity over 70%.	[70]
	GO	■ Fastest response reported of up to 0.18 s. ■ Recovery time of 0.3 s. ■ Operativity from 37% up to 98% relative humidity.	[71]
	Zinc oxide-tailored graphene foam	■ High stability. ■ High regenerability. ■ Linearity from 20% up to 95% relative humidity.	[72]
	Zinc oxide-tailored GO	■ Response of up to 1 s. ■ Linearity from 20% up to 95% relative humidity.	[73]
	Silver nanoparticle-tailored GO	■ Sensitivity of 26 nF/% RH. ■ Linear range from 11% up to 87% relative humidity.	[74]
	N-[4-morpholinecarboximidamidoyl] carboximidamidoylated GO	■ Response of up 20. ■ Recovery time of 2 s.	[75]
NH_3	Chemical vapor-deposited graphene	■ Properties related to graphene layer numbers.	[76]
	GO	■ Properties related to graphene layer numbers. ■ Linear range from 10 to 100 ppm. ■ Single-layer performed better than double- and multilayer electrodes.	[77]
	Fluorinated GO	■ Improvement over 7% performance compared with GO.	[78]
	Phosphorous-doped graphene	■ Limit of detection of 69 ppb. ■ Improvement over 70% of electrochemical performance compared with pristine graphene.	[79]
	Aniline-tailored graphene	■ Easy fabrication. ■ Response of 37% in 50 ppm of NH_3.	[80]
	Zinc oxide on rGO	■ Fast response up to 2 s. ■ Fast recovery up to 13 s in 350 ppm of NH_3. ■ Detection limit of 10 ppm.	[81]
	$CuFe_2O_4$-tailored rGO	■ Fast response up to 3 s. ■ Fast recovery up to 6 s. ■ High selectivity (over 5 times) for NH_3 in presence of CH_3OH, CO_2, benzene. ■ Limit of detection of 5 ppm.	[82]

Table 1. Cont.

Gas	Material	Highlights	References
NO$_2$	Multilayered porous graphene	■ Selective for NO$_2$ in presence of NH$_3$. ■ Detection limit of 25 ppb. ■ Response independent from relative humidity.	[83]
	Silicon-doped graphene	■ High response value of up to 22 in 50 ppm of NO$_2$. ■ Fast response up to 126 s. ■ Fast recovery up to 378 s. ■ Linear range from 18 ppb up to 300 ppm. ■ Good selectivity.	[84]
	Phosphorous-doped graphene	■ High response value of up to 59% in 50 ppm of NO$_2$. ■ Detection limit 1 ppm.	[85]
	Metal frameworks on rGO	■ Detection limit 0.7 ppm. ■ Selective for NO$_2$ in presence of NH$_3$. ■ Nonselective for NO$_2$ in presence of NO.	[86]
	Cobalt hydroxide-tailored rGO	■ High sensitivity of 70% exposed to 100 ppm of NO$_2$. ■ Detection limit of 1 ppm.	[87]
	Mixed iron and cobalt oxide-tailored graphene	■ Good response of up to 50 s. ■ Detection limit of 1 ppm.	[88]
	Copper nanoparticle-tailored graphene	■ Great reproducibility. ■ Detection limit of 30 ppb. ■ Slow response.	[89]
H$_2$S	Zinc oxide onto rGO	■ Poor selectivity in presence of NO. ■ Detection limit of 8 ppm.	[90]
	Tin oxide onto rGO	■ Fast response in 2 s using 50 ppm of H$_2$S. ■ Response up to 30%. ■ Regenerable.	[91]
	Cobaltite supported on graphene nanospheres	■ Response of 30% in presence of 50 ppm of H$_2$S. ■ Linear range from 1 to 70 ppm.	[92]
SO$_2$	Annealed rGO	■ Stability over 30 days in sulphur dioxide atmosphere. ■ Detection limit 5 ppm.	[93]
	Sheets of GO	■ Moderate response up to 65 s. ■ Fast recovery up to 100 s. ■ Detection limit of up to 15 ppm.	[94]
	rGO	■ Response of up to 47% in 50 ppm of SO$_2$. ■ Detection limit of up to 5 ppm.	[95]

GRMs showed some key advantages over other 2D materials such as MXenes, mostly focused on their preparation and tailoring. The synthesis of MXenes is a complex multistage process that should operate in well-controlled conditions [96] for the production of a high-quality material, similar to the single-layer graphene process. Nevertheless, GRM production has been developed and optimized for scalability, as proven by the production of GO and rGO from a wide range of cheap feedstocks under mild conditions through robust processes [97–99]. Furthermore, the carbonaceous low-dimensional materials can exploit a wide range of reactivity, fostering an easy chemical tailoring and preserving their properties [100].

3.1. Pristine Graphene and GRM Sensing Performance in Electrochemical Gas Detection: CO$_2$ and CO

Monitoring the asphyxiating gases produced from combustion, such as CO$_2$ and CO, is a relevant safety issue [101]. The main issue of detecting CO$_2$ through electrochemical sensing is the interference of other species in the atmosphere, such as water, CO and oxygen. Smith et al. [44] investigated the cross-sensitivity of a capacitive CO$_2$ sensor in the presence of several residual atmospheres (Ar, H$_2$O, N$_2$) using a chemical vapor-deposited single-layer graphene. Particularly, the authors investigated the effect of humidity on the sensor performance, showing the absence of sensitivity towards CO_2 in the presence of atmospheric-level humidity. The authors simulated through density functional theory (DFT) calculations the effect of H_2O and CO_2, showing that the reduction in sensitivity towards CO_2 was due to the electronic alteration of graphene induced by the adsorbed water molecule. The study of CO_2 with GRMs is of great interest for producing high-performance gas sensors, and it is affected by several key factors, such as doping, as reported in Figure 2.

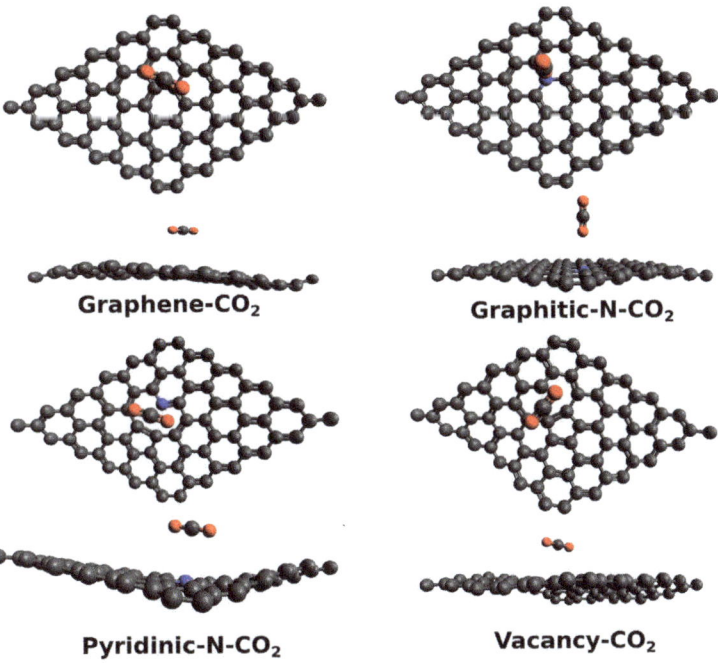

Figure 2. Computational simulation of the interaction between CO_2 and graphene or GRMs at 25 °C (carbon atoms were reported as black, oxygen atoms were reported as red and nitrogen ones as blue). Reprinted with all permission from del Castillo et al. [102].

Castillo et al. [102] investigated the role of heteroatom-doped graphene in sensing CO_2, proving that the presence of nitrogen graphitic sites can alter the local morphology of graphene sheets and improving the sensitivity towards CO_2 over that achievable by using a Pt-decorated electrode. The authors suggested that this was due to the very same nature of nitrogen graphitic sites that act as p-type doping agents. This induced a pullout of electrons from CO_2, improving the electrocatalytic activity of the nitrogen-doped graphene.

Deji et al. [103] evaluated the effect of boron and phosphorous co-doping of graphene nanoribbons for direct CO_2 detection using first-principle DFT simulation. The authors reported that phosphorous-doped graphene showed an adsorption energy eight times higher than pristine material, while the boron-doped one outperformed it. This study is of particular significance as it assesses the relevance of the doping agent. Additionally, Elgammal et al. [104] also proved that the support onto which graphene is deposited affects the sensing process, even if not in such a relevant way. The authors modelled the performance of graphene supported on silica or sapphire substrates for detecting both CO_2 and H_2O molecules using DFT simulations. The results showing the differences between the substrates are in the range of 1 to 10 meV. Interestingly, authors reported that H_2O molecules prefer to be adsorbed onto hollow sites in the center of the graphene hexagonal moieties, while CO_2 molecules prefer sites bridging carbon–carbon bonds or directly on top of carbon atoms. Also, the authors reported that the adsorption energy of CO_2 was up to 0.17 eV, while H_2O showed values close to 0.09 eV.

The weak interactions between CO_2 and graphene are a relevant issue, but several studies reported the possibility of using GRMs as solid materials for CO_2 sensing. Yoon et al. [45] assembled a CO_2 sensor fabricated by mechanical cleavage of nanographite plates. The authors were able to detect CO_2 at room temperature in the presence of water (humid conditions), observing a linear response of conductance in the range between 10 and 100 ppm. Fan et al. [46] used a double-layer impedimetric electrode for the detection of CO_2 without observing any significant influence of H_2O for relative humidity (RH) up to 3%. The authors also proved that double-layer graphene performed better than single-layer due to the different spatial distribution of the electronic density.

GRMs have also been diffusely used for improving the interaction with CO_2. Akhter et al. [47] designed a low-cost, low-power, miniature, highly sensitive and selective impedimetric CO_2 sensor using GO. The authors reported a linear range from 400 ppm to 4000 ppm with good performance in reproducibility and stability. Furthermore, they also achieved a very negligible cross-sensitivity with H_2O and a fast response and recovery rate. Muhammad Hafiz et al. [48] used rGO produced by hydrogen plasma as an impedimetric sensor. The authors reported a CO_2 gas-sensing response of 71% (calculated as resistance variation of the electrode) in the presence of a CO_2 concentration up to 1500 ppm in N_2 and 37% RH, while the performance decreased down to a response of 15% in air environment with 68% RH. Nevertheless, the sensor showed a fast response and a good recovery rate. Alternatively, GRMs that include metal species can be used, as reported by Miao et al. [49]. The authors developed a platform system able to operate from 10 up to 60 °C with 97% RH and a CO_2 linear response ranging from 300 up to 1100 ppm.

CO showed a different interaction geometry with graphene, as reported by the computational study of Dindorkar et al. [105] and summarized in Figure 3.

Figure 3. Computational simulation of the interaction between CO and graphene at 25 °C (carbon atoms were reported as black, oxygen atoms were reported as red and hydrogen ones as white). Reprinted with all permission from Dindorkar et al. [105].

While CO_2 preferentially interacts with carbon atoms, the CO preferential interaction is with the edges of graphene sheets (Figure 2). The authors also found that CO interacts directly with carbon atoms but only in highly doped fragments containing silicon carbide or boron nitride domains. Similar effects were reported in the presence of GO by Deji and co-workers [106] that also proved the effectiveness of the tailoring process, with metal nanoparticles decreasing the adsorption energy up to 40 times compared with pristine graphene [107]. Metal oxides combined with graphene and GRMs are able to form p-n junctions, increasing the conductivity and improving sensing performance [108]. When it comes to CO sensing, Pd has shown the most remarkable performance, as reported by Kashyap et al. [50], using palladium-tailored rGO as an impedimetric sensor. The authors tested the response to CO in the presence of both CH_4 and H_2, suggesting that the interaction mechanism of CO was lying between the sole interactions with electron withdrawing or electron donating. As reported by Shojaee et al. [51], the combination of Pd with a metal oxide such as SnO_2 could be particularly beneficial for both response and recovery, improving, at the same time, the specific surface area of the material. The authors also provided an overview of the effect of RH on a 400 ppm CO sample analysis, reporting a decrement of response for RH up to 60%. They ascribed this behavior to the competitive adsorption of CO and water molecules on the Pd and SnO_2. A further RH increment of up to 85% induced a considerable increment of electrode response due to the formation of surface conductive channels [67]. The surface porosity was also investigated by Ha et al. [52] using ZnO nanoparticles onto rGO. The authors achieved an electrode response value of 85% for 1000 ppm CO at 200 °C, with a recovery time of 9 s. Similarly, the response value, response time, and recovery time of the sensor at room temperature were 27.5%, 14 s, and 15 s, respectively. The sensor demonstrated a distinct response to various CO concentrations in the range of 1–1000 ppm and good selectivity towards CO gas. In addition, the sensor exhibited good repeatability in multicycle and long-term stability. Neetha et al. [53] decorated rGO with Mn_3O_4, achieving a response time of only 3 s at 25 °C using 50 ppm of CO. Similar results were obtained using SnO_2 on graphene [54], CuO on rGO [56] and mixed metal oxide over rGO [55,57].

Nevertheless, inorganic tailoring is not a mandatory condition for detecting CO. iGO by itself can act as an active material for the detection of CO, as reported by Panda et al. [58]. The authors achieved a 71% sensitivity using 30 ppm CO at room temperature (RT), with a recovery time of up to 30 s and a remarkable sensitivity of up to 10 ppm. The authors suggested that the performance was due to the in situ production of atomic, ionic, and radical oxygen sites, which play a relevant role in both the adsorption of CO and electronic density rearmament. Furthermore, GRMs can be functionalized with polymers, as reported by Farea and co-workers [59,60] and by Mohammed et al. [61], or by metal organic frameworks boosting the overall CO sensing performance, as reported by More et al. [62].

3.2. Pristine Graphene and GRM Sensing Performance in Electrochemical Gas Detection: H_2

H_2 is among the most elusive gases to be detected, and neat graphene cannot be used for direct sensing of it. The most common strategy for H_2 sensing is tailoring the GRM surface with metal nanoparticles, activating a mechanism known as spillover, as reported in Figure 4.

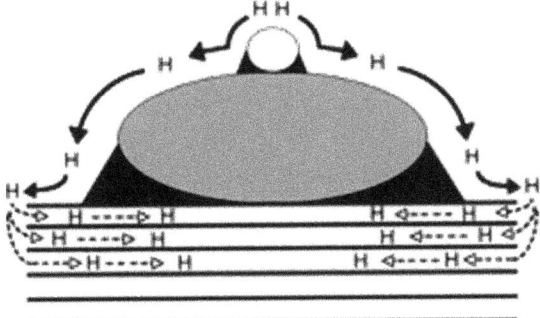

Figure 4. Schematic representation of primary and secondary spillover promoted by the presence of metal nanoparticles and GRMs. Reprinted with all permission from Lachawiec [109] (*Copyright © 2005, American Chemical Society*).

H_2 spillover is a complex phenomenon occurring when H_2 molecules dissociate onto a metal particle and diffuse as atomic hydrogen to the graphene support, while the second spillover involves a further transportation towards the carbon support. This behavior can be modulated by the introduction of layered GRMs, as reported by Kumar et al. [110], who modelled H_2 sensing in a GRM containing a layer of antimonene. DFT calculations showed the presence of a Bader charge transfer mechanism from the antimonene layer towards the graphene one that was able to change the potential barrier from the Ohmic to the Schottky type. Moving to tailored GRMs, Pd and Pt are the higher-performing metals due to their ability to interact with hydrogen through adsorption and release processes [111–118]. As reported by Kishnani et al. [119], palladium-doped or -decorated graphene is very effective in sensing H_2. Particularly, the authors reported a higher electrochemical activity and conductivity for palladium-decorated graphene compared with the palladium-doped one, while the charge transfer and recovery time showed an opposite trend. Chung et al. [63] decorated a single-layer graphene sheet with palladium nanoparticles of 3 nm of average size. The authors reported a response of 33% using 1000 ppm of H_2 at 25 °C and a remarkable detection limit of 20 ppm. The effect of palladium active centers was also observed by Lange et al. [120] using cyclic voltammetry without providing any highlights on the mechanism. Lee et al. [64] incorporated palladium nanoparticles into a 3D-GRM structure, reaching a response of 41.9% under 3% H_2 residual atmosphere.

Platinum has also been investigated as a viable alternative to palladium. Lu et al. [65] decorated rGO with platinum nanoparticles by using freeze-drying-assisted techniques, reaching a sensitivity toward 0.5% hydrogen up to 8% and a recovery time of 63 s. Similarly, Lee et al. [64] and Phan and co-workers [66] produced a highly porous 3D-GRM containing platinum nanoparticles, achieving good linearity from 1 to 100 ppm. As reported by Russo et al. [67], the addition of SnO_2 to palladium-decorated rGO was particularly beneficial, enhancing the response of palladium-decorated GRMs over four times. The authors suggested that the enhancement of sensing performance was due to the formation of a heterojunction between the n-type SnO_2 and the p-type rGO in the heterostructure, boosting the catalytic effect of platinum in promoting the dissociation of H_2.

Non-noble metal oxides have also been used extensively coupled with GRMs for H_2 detection. Ahmad Fauzi et al. [68] decorated a proton-conducting GO membrane with WO_3, producing a potentiometric H_2 sensor. The authors claimed a response of 50 mV in the presence of H_2 100 ppm in air atmosphere and a detection limit of 11 ppm. ZnO was also used with GO, with interesting results, as reported by Rasch et al. [69]. The authors achieved a very low detection limit of 4 ppm, with a very fast response of around 114 s and a small recovery time of 30 s.

3.3. Pristinine Graphene and GRM Sensing Performance in Electrochemical Gas Detection: H_2O

Humidity sensors play a critical role in several industrial sectors, such as semiconductor production, in which moisture content in the air is a critical parameter [121]. GRMs provide interesting solutions to detecting the moisture content of air, even at low concentrations, due to the interaction occurring between the graphene surface and H_2O molecules, as sketched in Figure 5.

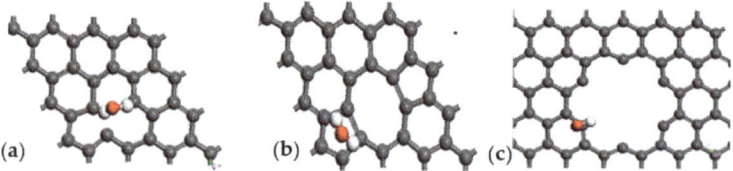

Figure 5. Computational simulation of the interaction between H_2O and graphene at 25 °C in (**a**) vacancy defect, (**b**) 5–7 defect and (**c**) close to hole defect. Reprinted with all permission from del Ye et al. [122].

As reported in Figure 3, H_2O molecules interact with graphene sheets without bonding and only through weak interactions with a distance of 2.5 Å and a higher deformation due to the interaction close to hole defects (Figure 5c), while the other cases (Figure 3a,b) did not show any significant distortion. Wang et al. [70] utilized vertically aligned graphene arrays as a humidity sensor platform. The authors observed the rise of system current with the increment of RH, suggesting a link to the Schottky barrier height with the junction resistance decrement due to the adsorption of vapor molecules. The authors suggested that the water molecules act as electron acceptors, increasing the hole density in the graphene systems. Zeng et al. [71] produced a self-powered H_2O flexible sensor with ultrafast response and recovery time of up to 0.3 s using GO. The authors claimed to have obtained a faster response in the field of humidity sensors, and they were able to operate in an RH range from 33 to 98%. Interestingly, they proposed an interpretative model of H⁺ diffusion occurring during the sensing process based on Grotthuss hopping [123]. Yu et al. [121] produced a sensor based on rGO with the highest sensitivity for humidity. The authors suggested that this exceptional behavior was due to the spherical double surfaces and small pores in the 3D structure of rGO, allowing an optimal exposure of functionalities and a magnification of H_2O interactions. Nevertheless, both GO and rGO suffer several issues, such as cross-reactivity towards the other gas present in the analyte. Seeneevassen et al. [124] used a GO sensor to monitor the humidity in an effluent gas, observing that the electrode must be conditioned before use through several cycles of humidification–dehumidification.

Huang et al. [125] faced the problem of humidity quantification in the agricultural sector, in which there are many interfering agents. The authors encapsulated rGO under a layer of GO deposited by spray coating, observing a reduction in cross-sensitivity towards both NH_3 and ethanol and retaining a good response and high sensitivity of up to 0.4% RH. As for the detection of other species, GRM tailoring significantly helps the sensing process, as proven by decoration with metal oxides [70,72], metal nanoparticles or organic fragments [75].

3.4. Pristine Graphene and GRM Sensing Performance in Electrochemical Gas Detection: NH_3 and NO_x

The detection of NH_3 is highly interesting due to the harmfulness of this gas [126,127]. The interaction with graphene is also, in this case, the key to understanding how to optimize the NH_3 sensing process. As shown in Figure 6 [128], NH_3 interaction with graphene occurs mainly through weak hybridizations between graphene and NH_3 p orbitals. Accordingly, NH_3 acts as an electron donor, but pristine graphene is not able to promote both good adsorption and an efficient small transfer charge, resulting in poor detection ability.

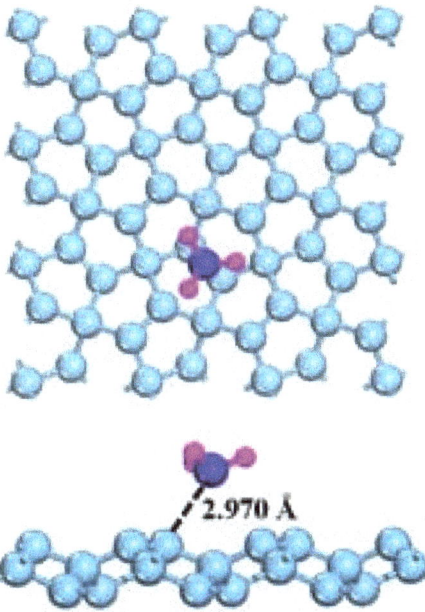

Figure 6. Computational simulation of the interaction between NH_3 and graphene at 25 °C (carbon atoms were reported as light blue, nitrogen atoms were reported as dark blue and hydrogen ones as purple). Reprinted with all permission from Chen et al. [128].

The interaction between NH_3 and graphene was evaluated by Song et al. [76]. The authors investigated the connection between sensitivity and graphene layers using single-layer, double-layer and multilayer graphene and 12,500 ppm of NH_3. The results showed that the electron transfer is three times higher in single-layer graphene than in the other species due to the easy rearrangement of charge density. Su et al. [77] used a similar approach for the production of a GO layered sensor. Also, in this case, the single-layer material showed the best performance, with a linear range from 5 to 100 ppm and a sensitivity over 15% greater than the multilayered GO. Among GRMs, fluorinated graphene also showed remarkable performance for NH_3 sensing, showing a 7% change in the resistive response, while pristine GO did not show any sensing ability [78]. This behavior was due to the lower Fermi level of GO and to the increment of hole density in fluorinated GO. Alternatively, Li et al. [79] doped graphene using phosphorous for NH_3 sensing. The authors reported an increment in performance with a reduction in both response and recovery time up to 71% and 73%, respectively and a detection limit of 69 ppb. Furthermore, the phosphorous-doped system showed a remarkable combination of repeatability, stability, and selectivity. The functionalization using both organic [80] and inorganic [81,82] species is also, in this case, a powerful tool for enhancing the electrochemical properties of GRMs.

NO_x species represent the other great family of hazardous nitrogen-based gases [129]. Matatagui et al. [83] faced the issue represented by the detection of NO_2 in the presence of NH_3 using a multilayered porous graphene electrode. After photoactivation, the authors reported a change in resistance of 16% in the presence of 0.5 ppm NO_2, with a detection limit of around 25 ppb. Interestingly, the response to both NH_3 (50 ppm) and H_2O (TH 33%) is negligible. The graphene decoration allowed for further improvement in the performance of the sensor by including silicon [84] or phosphorus [85], reaching a response of up to 22% and 59% resistance change using 50 ppm of NO_2.

Duy et al. [86] deeply explored the tailoring of the rGO surface with metal frameworks, including TiO_2 nanoparticles and WO_3, WS_2, and MoS_2 nanoflakes using cellulose as a binder. The authors reported a sensitivity boost towards NO_2, improving the detection limit up to 0.7 ppm using MoS_2 nanoflakes. ZnO oxide nanoparticles have also been diffusely used for the same scope [130–132] with poor results if compared with other nanostructures, such as cobalt supported on rGO [87], chromium-tailored graphene [133], and mixed iron and cobalt oxide [88] or copper [89] onto graphene.

3.5. Pristine Graphene and GRM Sensing Performance in Electrochemical Gas Detection: H_2S and SO_2

As previously described for NH_3, the interaction between GRMs and H_2S or SO_2 occurs mainly by p-orbital interactions with the π graphene system. As shown in Figure 7a,b, H_2S or SO_2 interact with different geometry and distances due to the differences in the electron acceptor behavior of SO_2, while H_2S acts as an electron donor. Nevertheless, pristine graphene is only poorly able to detect them as a consequence of very weak interactions with them [134].

As reported in the computational research carried out by Liu et al. [87], the insertion of a dopant agent such as aluminum atoms together with Stone–Wales defects can be beneficial for boosting the adsorption of SO_2, and a similar result was reported for H_2S [135].

Ugale et al. [90] approached the detection of H_2S using ZnO-tailored rGO fibers, reaching a detection limit of 8 ppm but a poor selectivity in the presence of NO. Song et al. [91] obtained better performance using SnO_2 supported on rGO, achieving a 33% response in 2 s using 50 ppm of H_2S. Furthermore, this system was totally reversible at 22 °C, allowing for long-time use. Similar results were obtained by using Co_3O_4-tailored graphene nanospheres [92] or by using copper or WO_3 supported on rGO [92]. Even if the H_2S and SO_2 sensing using GRMs is a field of great interest for both health and safety, the majority of the published research is currently focused more on the computational point of view rather than the applicative process, creating a perilous gap in the research [135].

It is noteworthy that Kumar et al. [93,95] deeply investigated the utilization of rGO for the detection of SO_2, achieving a limit of detection of up to 5 ppm by using annealed rGO.

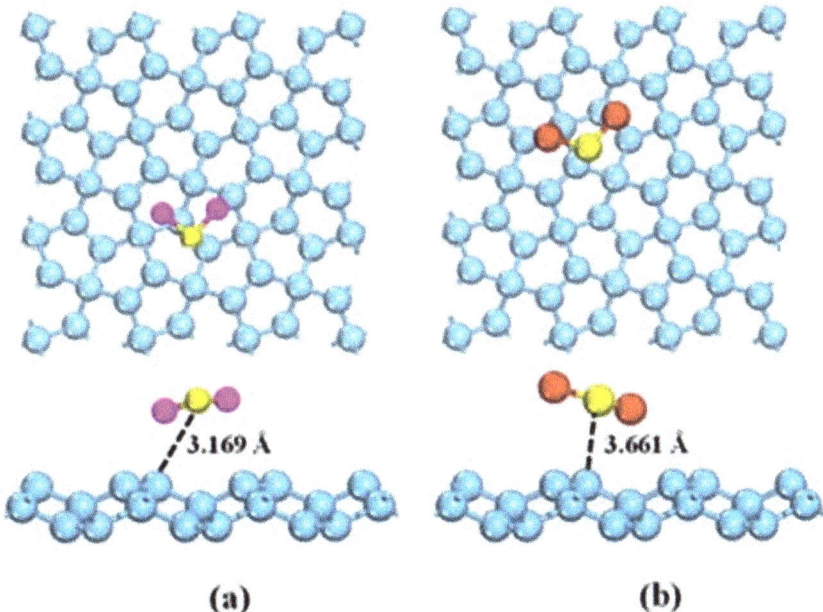

Figure 7. Computational simulation of the interaction between (**a**) H_2S and (**b**) SO_2 and graphene at 25 °C(carbon atoms were reported as light blue, sulphur atoms were reported as yellow, oxygen atoms were reported as red and hydrogen ones as purple. Reprinted with all permission from del Chen et al. [128].

3.6. Future Outlook for Pristine Graphene and GRM Gas Sensors: Wearable Devices

GRM-based gas sensors are still far from being affordable, but they show interesting perspectives for application in the production of wearable devices [136]. Wearable GRM gas sensors combine flexibility and light weight, enabling the creation of sensors integrated into clothing or into accessories and providing a non-intrusive solution for continuous gas monitoring [137,138].

This sensor family will allow for real-time monitoring, enabling continuous tracking of environmental and personal exposure to gases [139] and promoting healthcare for several activities, including those in which workers can be exposed to hazardous gases. As reported in Figure 8, Peng et al. [140] produced a humidity sensor based on laser-induced graphene that is stretchable and able to operate in real industrial environments.

The wearable GRM gas sensor has another key feature, i.e., high energy efficiency, that allows for prolonged usage without the need for frequent recharging [141]. As shown in Figure 9, Sun et al. produced a GRM-based platform able to exploit several functions, such as real-time monitoring of temperature, hydration and sweat due to the remarkable water vapor permeability.

Monitoring using high-performance electrodes will allow a significant improvement in the safety of operation in vulnerable environments such as the semiconductor and food industries, where the atmosphere should be continuously monitored, and for chemical industries operating with complex gas mixtures at high temperatures and pressures.

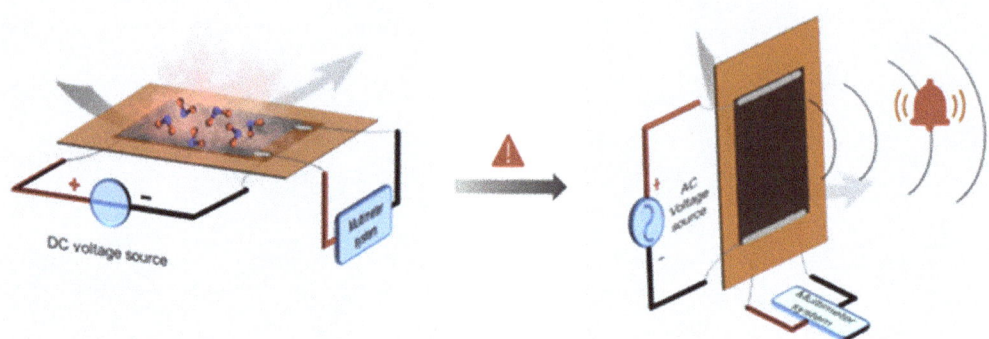

Figure 8. Stretchable laser-induced graphene-based humidity sensor. Reprinted with all permission from del Peng et al. [140].

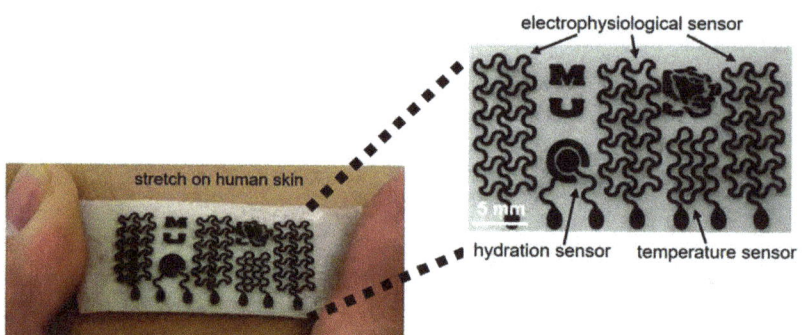

Figure 9. Biocompatible gas sensors based on GRM millimetric circuits. Reprinted with all permission from del Sun et al. [142].

4. Conclusions

The field of gas sensors is of paramount relevance for both health and safety. The development of new high-performance materials is the key to the future of the field, and GRMs can play a relevant and active role. Nowadays, their superior performance is counterbalanced by their high cost, but a great effort has been devoted to making their industrial production more economically feasible. Nevertheless, the usage of GRM-based gas sensors can easily reach all those applications in which performance is more important than economics. Furthermore, the field of wearable gas sensors for monitoring both people's metabolism and the surrounding environment is a field of application where GRMs represent the state of the art.

We believe in the near-future scenario in which these materials will reach and improve other key sectors of daily life.

Author Contributions: Conceptualization, M.B. and A.T.; writing—original draft preparation, M.M. and M.B.; writing—review and editing, M.M., M.B. and A.T.; visualization, M.M. and M.B.; supervision, M.B. and A.T. All authors have read and agreed to the published version of the manuscript.

Funding: This research received no external funding.

Conflicts of Interest: The authors declare no conflicts of interest.

References

1. Baranwal, J.; Barse, B.; Gatto, G.; Broncova, G.; Kumar, A. Electrochemical sensors and their applications: A review. *Chemosensors* **2022**, *10*, 363. [CrossRef]
2. Fu, L.; Mao, S.; Chen, F.; Zhao, S.; Su, W.; Lai, G.; Yu, A.; Lin, C.-T. Graphene-based electrochemical sensors for antibiotic detection in water, food and soil: A scientometric analysis in CiteSpace (2011–2021). *Chemosphere* **2022**, *297*, 134127. [CrossRef] [PubMed]
3. Benjamin, S.R.; Junior, E.J.M.R. Graphene based electrochemical sensors for detection of environmental pollutants. *Curr. Opin. Environ. Sci. Health* **2022**, *29*, 100381. [CrossRef]
4. Kanjwal, M.A.; Ghaferi, A.A. Graphene incorporated electrospun nanofiber for electrochemical sensing and biomedical applications: A critical review. *Sensors* **2022**, *22*, 8661. [CrossRef] [PubMed]
5. Hashim, N.; Abdullah, S.; Yusoh, K. Graphene nanomaterials in the food industries: Quality control in promising food safety to consumers. *Graphene 2D Mater.* **2022**, *7*, 1–29. [CrossRef]
6. Han, Z.; Zhang, X.; Yuan, H.; Li, Z.; Li, G.; Zhang, H.; Tan, Y. Graphene oxide/gold nanoparticle/graphite fiber microelectrodes for directing electron transfer of glucose oxidase and glucose detection. *J. Power Sources* **2022**, *521*, 230956. [CrossRef]
7. Yue, J.; Li, C.; Ji, X.; Tao, Y.; Lu, J.; Cheng, Y.; Du, J.; Wang, H. Highly tough and conductive hydrogel based on defect-patched reduction graphene oxide for high-performance self-powered flexible sensing micro-system. *Chem. Eng. J.* **2023**, *466*, 143358. [CrossRef]
8. Panraksa, Y.; Siangproh, W.; Khampieng, T.; Chailapakul, O.; Apilux, A. based amperometric sensor for determination of acetylcholinesterase using screen-printed graphene electrode. *Talanta* **2018**, *178*, 1017–1023. [CrossRef]
9. Jiang, Y.; Zhang, X.; Shan, C.; Hua, S.; Zhang, Q.; Bai, X.; Dan, L.; Niu, L. Functionalization of graphene with electrodeposited Prussian blue towards amperometric sensing application. *Talanta* **2011**, *85*, 76–81. [CrossRef]
10. Salvo, P.; Melai, B.; Calisi, N.; Paoletti, C.; Bellagambi, F.; Kirchhain, A.; Trivella, M.G.; Fuoco, R.; Di Francesco, F. Graphene-based devices for measuring pH. *Sens. Actuators B Chem.* **2018**, *256*, 976–991. [CrossRef]
11. Jaworska, E.; Lewandowski, W.; Mieczkowski, J.; Maksymiuk, K.; Michalska, A. Critical assessment of graphene as ion-to-electron transducer for all-solid-state potentiometric sensors. *Talanta* **2012**, *97*, 414–419. [CrossRef] [PubMed]
12. Bonanni, A.; Loo, A.H.; Pumera, M. Graphene for impedimetric biosensing. *TrAC Trends Anal. Chem.* **2012**, *37*, 12–21. [CrossRef]
13. Marć, M.; Tobiszewski, M.; Zabiegała, B.; de la Guardia, M.; Namieśnik, J. Current air quality analytics and monitoring: A review. *Anal. Chim. Acta* **2015**, *853*, 116–126. [CrossRef] [PubMed]
14. Pitarma, R.; Marques, G.; Ferreira, B.R. Monitoring indoor air quality for enhanced occupational health. *J. Med. Syst.* **2017**, *41*, 23. [CrossRef] [PubMed]
15. Basu, S.; Bhattacharyya, P. Recent developments on graphene and graphene oxide based solid state gas sensors. *Sens. Actuators B Chem.* **2012**, *173*, 1–21. [CrossRef]
16. International Union of Pure and Applied Chemistry (IUPAC). Graphene Layer, 3.0.1 ed. Available online: https://goldbook.iupac.org/terms/view/G02683 (accessed on 4 January 2024).
17. Mintmire, J.W.; Dunlap, B.I.; White, C.T. Are fullerene tubules metallic? *Phys. Rev. Lett.* **1992**, *68*, 631. [CrossRef]
18. Yan, J.-A.; Ruan, W.; Chou, M. Electron-phonon interactions for optical-phonon modes in few-layer graphene: First-principles calculations. *Phys. Rev. B* **2009**, *79*, 115443. [CrossRef]
19. Dresselhaus, M.; Jorio, A.; Saito, R. Characterizing graphene, graphite, and carbon nanotubes by Raman spectroscopy. *Annu. Rev. Condens. Matter Phys.* **2010**, *1*, 89–108. [CrossRef]
20. Liu, C.-C.; Walters, A.B.; Vannice, M.A. Measurement of electrical properties of a carbon black. *Carbon* **1995**, *33*, 1699–1708. [CrossRef]
21. Lavagna, L.; Meligrana, G.; Gerbaldi, C.; Tagliaferro, A.; Bartoli, M. Graphene and Lithium-Based Battery Electrodes: A Review of Recent Literature. *Energies* **2020**, *13*, 4867. [CrossRef]
22. Siegel, D.A.; Regan, W.; Fedorov, A.V.; Zettl, A.; Lanzara, A. Charge-carrier screening in single-layer graphene. *Phys. Rev. Lett.* **2013**, *110*, 146802. [CrossRef] [PubMed]
23. Jiang, Z.; Zhang, Y.; Tan, Y.-W.; Stormer, H.; Kim, P. Quantum Hall effect in graphene. *Solid State Commun.* **2007**, *143*, 14–19. [CrossRef]
24. Xu, Z.; Zheng, Q.-S.; Chen, G. Elementary building blocks of graphene-nanoribbon-based electronic devices. *Appl. Phys. Lett.* **2007**, *90*, 223115. [CrossRef]
25. Neto, A.C.; Guinea, F.; Peres, N.M.; Novoselov, K.S.; Geim, A.K. The electronic properties of graphene. *Rev. Mod. Phys.* **2009**, *81*, 109. [CrossRef]
26. Jariwala, D.; Srivastava, A.; Ajayan, P.M. Graphene synthesis and band gap opening. *J. Nanosci. Nanotechnol.* **2011**, *11*, 6621–6641. [CrossRef] [PubMed]
27. Han, M.Y.; Özyilmaz, B.; Zhang, Y.; Kim, P. Energy band-gap engineering of graphene nanoribbons. *Phys. Rev. Lett.* **2007**, *98*, 206805. [CrossRef]
28. Li, X.; Wang, X.L.; Zhang, L.; Lee, S.; Dai, H. Chemically Derived, Ultrasmooth Graphene Nanoribbon Semiconductors. *Science* **2008**, *319*, 1229. [CrossRef]
29. Rani, P.; Jindal, V. Designing band gap of graphene by B and N dopant atoms. *RSC Adv.* **2013**, *3*, 802–812. [CrossRef]
30. Lerf, A.; He, H.; Forster, M.; Klinowski, J. Structure of graphite oxide revisited. *J. Phys. Chem. B* **1998**, *102*, 4477–4482. [CrossRef]

31. Krishnamoorthy, K.; Veerapandian, M.; Yun, K.; Kim, S.-J. The chemical and structural analysis of graphene oxide with different degrees of oxidation. *Carbon* **2013**, *53*, 38–49. [CrossRef]
32. Zhu, Y.; Murali, S.; Cai, W.; Li, X.; Suk, J.W.; Potts, J.R.; Ruoff, R.S. Graphene and graphene oxide: Synthesis, properties, and applications. *Adv. Mater.* **2010**, *22*, 3906–3924. [CrossRef] [PubMed]
33. Hunt, A.; Kurmaev, E.; Moewes, A. Band gap engineering of graphene oxide by chemical modification. *Carbon* **2014**, *75*, 366–371. [CrossRef]
34. Hernández Rosas, J.; Ramírez Gutiérrez, R.; Escobedo-Morales, A.; Chigo Anota, E. First principles calculations of the electronic and chemical properties of graphene, graphane, and graphene oxide. *J. Mol. Model.* **2011**, *17*, 1133–1139. [CrossRef] [PubMed]
35. Zhu, Y.; James, D.K.; Tour, J.M. New routes to graphene, graphene oxide and their related applications. *Adv. Mater.* **2012**, *24*, 4924–4955. [CrossRef] [PubMed]
36. Guex, L.G.; Sacchi, B.; Peuvot, K.F.; Andersson, R.L.; Pourrahimi, A.M.; Ström, V.; Farris, S.; Olsson, R.T. Experimental review: Chemical reduction of graphene oxide (GO) to reduced graphene oxide (rGO) by aqueous chemistry. *Nanoscale* **2017**, *9*, 9562–9571. [CrossRef]
37. Ahmed, A.; Singh, A.; Young, S.-J.; Gupta, V.; Singh, M.; Arya, S. Synthesis techniques and advances in sensing applications of reduced graphene oxide (rGO) composites: A review. *Compos. Part A Appl. Sci. Manuf.* **2022**, *165*, 107373. [CrossRef]
38. Wang, Y.; Chen, Y.; Lacey, S.D.; Xu, L.; Xie, H.; Li, T.; Danner, V.A.; Hu, L. Reduced graphene oxide film with record-high conductivity and mobility. *Mater. Today* **2018**, *21*, 186–192. [CrossRef]
39. Mouhat, F.; Coudert, F.-X.; Bocquet, M.-L. Structure and chemistry of graphene oxide in liquid water from first principles. *Nat. Commun.* **2020**, *11*, 1566. [CrossRef]
40. Zhao, G.; Li, X.; Huang, M.; Zhen, Z.; Zhong, Y.; Chen, Q.; Zhao, X.; He, Y.; Hu, R.; Yang, T. The physics and chemistry of graphene-on-surfaces. *Chem. Soc. Rev.* **2017**, *46*, 4417–4449. [CrossRef]
41. Catania, F.; Marras, E.; Giorcelli, M.; Jagdale, P.; Lavagna, L.; Tagliaferro, A.; Bartoli, M. A Review on Recent Advancements of Graphene and Graphene-Related Materials in Biological Applications. *Appl. Sci.* **2021**, *11*, 614. [CrossRef]
42. Szczęśniak, B.; Choma, J.; Jaroniec, M. Gas adsorption properties of graphene-based materials. *Adv. Colloid Interface Sci.* **2017**, *243*, 46–59. [CrossRef] [PubMed]
43. Huang, B.; Li, Z.; Liu, Z.; Zhou, G.; Hao, S.; Wu, J.; Gu, B.-L.; Duan, W. Adsorption of gas molecules on graphene nanoribbons and its implication for nanoscale molecule sensor. *J. Phys. Chem. C* **2008**, *112*, 13442–13446. [CrossRef]
44. Smith, A.D.; Elgammal, K.; Fan, X.; Lemme, M.C.; Delin, A.; Råsander, M.; Bergqvist, L.; Schröder, S.; Fischer, A.C.; Niklaus, F.; et al. Graphene-based CO_2 sensing and its cross-sensitivity with humidity. *RSC Adv.* **2017**, *7*, 22329–22339. [CrossRef]
45. Yoon, H.J.; Jun, D.H.; Yang, J.H.; Zhou, Z.; Yang, S.S.; Cheng, M.M.-C. Carbon dioxide gas sensor using a graphene sheet. *Sens. Actuators B Chem.* **2011**, *157*, 310–313. [CrossRef]
46. Fan, X.; Elgammal, K.; Smith, A.D.; Östling, M.; Delin, A.; Lemme, M.C.; Niklaus, F. Humidity and CO_2 gas sensing properties of double-layer graphene. *Carbon* **2018**, *127*, 576–587. [CrossRef]
47. Akhter, F.; Alahi, M.E.E.; Siddiquei, H.R.; Gooneratne, C.P.; Mukhopadhyay, S.C. Graphene oxide (GO) coated impedimetric gas sensor for selective detection of carbon dioxide (CO_2) with temperature and humidity compensation. *IEEE Sens. J.* **2020**, *21*, 4241–4249. [CrossRef]
48. Muhammad Hafiz, S.; Ritikos, R.; Whitcher, T.J.; Razib, N.M.; Bien, D.C.S.; Chanlek, N.; Nakajima, H.; Saisopa, T.; Songsiriritthigul, P.; Huang, N.M.; et al. A practical carbon dioxide gas sensor using room-temperature hydrogen plasma reduced graphene oxide. *Sens. Actuators B Chem.* **2014**, *193*, 692–700. [CrossRef]
49. Miao, F.; Han, Y.; Shi, J.; Tao, B.; Zhang, P.; Chu, P.K. Design of graphene-based multi-parameter sensors. *J. Mater. Res. Technol.* **2023**, *22*, 3156–3169. [CrossRef]
50. Kashyap, A.; Barman, P.B.; Hazra, S.K. Low-temperature selectivity study of chemically treated graphene oxide for detection of hydrogen gas. *Mater. Today Proc.* **2023**; in press. [CrossRef]
51. Shojaee, M.; Nasrefahani, S.; Sheikhi, M.H. Hydrothermally synthesized Pd-loaded SnO_2/partially reduced graphene oxide nanocomposite for effective detection of carbon monoxide at room temperature. *Sens. Actuators B Chem.* **2018**, *254*, 457–467. [CrossRef]
52. Ha, N.H.; Thinh, D.D.; Huong, N.T.; Phuong, N.H.; Thach, P.D.; Hong, H.S. Fast response of carbon monoxide gas sensors using a highly porous network of ZnO nanoparticles decorated on 3D reduced graphene oxide. *Appl. Surf. Sci.* **2018**, *434*, 1048–1054. [CrossRef]
53. Neetha, J.; Abraham, K.E. Enhancement in carbon monoxide sensing performance by reduced graphene oxide/trimanganese tetraoxide system. *Sens. Actuators B Chem.* **2020**, *325*, 128749. [CrossRef]
54. Li, L.; Wang, C.; Ying, Z.; Wu, W.; Hu, Y.; Yang, W.; Xuan, W.; Li, Y.; Wen, F. SnO_2/graphene nanocomposite for effective detection of CO at room temperature. *Chem. Phys. Lett.* **2023**, *830*, 140803. [CrossRef]
55. Nandi, D.; Parameswaranpillai, J.; Siengchin, S. Mechanistic insight into high response of carbon monoxide gas sensor developed by nickel manganate nanorod decorated reduced graphene oxide. *Colloids Surf. A Physicochem. Eng. Asp.* **2020**, *589*, 124449. [CrossRef]
56. Zhang, D.; Jiang, C.; Liu, J.; Cao, Y. Carbon monoxide gas sensing at room temperature using copper oxide-decorated graphene hybrid nanocomposite prepared by layer-by-layer self-assembly. *Sens. Actuators B Chem.* **2017**, *247*, 875–882. [CrossRef]

57. Zhong, Y.; Li, M.; Tan, R.; Xiao, X.; Hu, Y.; Li, G. Co(III) doped-CoFe layered double hydroxide growth with graphene oxide as cataluminescence catalyst for detection of carbon monoxide. *Sens. Actuators B Chem.* **2021**, *347*, 130600. [CrossRef]
58. Panda, D.; Nandi, A.; Datta, S.K.; Saha, H.; Majumdar, S. Selective detection of carbon monoxide (CO) gas by reduced graphene oxide (rGO) at room temperature. *RSC Adv.* **2016**, *6*, 47337–47348. [CrossRef]
59. Farea, M.A.; Mohammed, H.Y.; Shirsat, S.M.; Ali, Z.M.; Tsai, M.-L.; Yahia, I.S.; Zahran, H.Y.; Shirsat, M.D. Impact of reduced graphene oxide on the sensing performance of Poly (3, 4–ethylenedioxythiophene) towards highly sensitive and selective CO sensor: A comprehensive study. *Synth. Met.* **2022**, *291*, 117166. [CrossRef]
60. Farea, M.A.; Mohammed, H.Y.; Shirsat, S.M.; Tsai, M.-L.; Murshed, M.N.; El Sayed, M.E.; Naji, S.; Samir, A.; Alsharabi, R.M.; Shirsat, M.D. A novel approach for ultrafast and highly sensitive carbon monoxide gas sensor based on PEDOT/GO nanocomposite. *Mater. Sci. Semicond. Process.* **2023**, *155*, 107255. [CrossRef]
61. Mohammed, H.Y.; Farea, M.A.; Ali, Z.M.; Shirsat, S.M.; Tsai, M.-L.; Shirsat, M.D. Poly(N-methyl pyrrole) decorated rGO nanocomposite: A novel ultrasensitive and selective carbon monoxide sensor. *Chem. Eng. J.* **2022**, *441*, 136010. [CrossRef]
62. More, M.S.; Bodkhe, G.A.; Ingle, N.N.; Singh, F.; Tsai, M.-L.; Kim, M.; Shirsat, M.D. Metal-organic framework (MOF)/reduced graphene oxide (rGO) composite for high performance CO sensor. *Solid-State Electron.* **2023**, *204*, 108630. [CrossRef]
63. Chung, M.G.; Kim, D.-H.; Seo, D.K.; Kim, T.; Im, H.U.; Lee, H.M.; Yoo, J.-B.; Hong, S.-H.; Kang, T.J.; Kim, Y.H. Flexible hydrogen sensors using graphene with palladium nanoparticle decoration. *Sens. Actuators B Chem.* **2012**, *169*, 387–392. [CrossRef]
64. Lee, B.; Cho, S.; Jeong, B.J.; Lee, S.H.; Kim, D.; Kim, S.H.; Park, J.-H.; Yu, H.K.; Choi, J.-Y. Highly responsive hydrogen sensor based on Pd nanoparticle-decorated transfer-free 3D graphene. *Sens. Actuators B Chem.* **2024**, *401*, 134913. [CrossRef]
65. Lu, X.; Song, X.; Gu, C.; Ren, H.; Sun, Y.; Huang, J. Freeze drying-assisted synthesis of Pt@reduced graphene oxide nanocomposites as excellent hydrogen sensor. *J. Phys. Chem. Solids* **2018**, *116*, 324–330. [CrossRef]
66. Phan, D.-T.; Youn, J.-S.; Jeon, K.-J. High-sensitivity and fast-response hydrogen sensor for safety application using Pt nanoparticle-decorated 3D graphene. *Renew. Energy* **2019**, *144*, 167–171. [CrossRef]
67. Russo, P.A.; Donato, N.; Leonardi, S.G.; Baek, S.; Conte, D.E.; Neri, G.; Pinna, N. Room-temperature hydrogen sensing with heteronanostructures based on reduced graphene oxide and tin oxide. *Angew. Chem. Int. Ed.* **2012**, *51*, 11053–11057. [CrossRef] [PubMed]
68. Ahmad Fauzi, A.S.; Hamidah, N.L.; Sato, S.; Shintani, M.; Putri, G.K.; Kitamura, S.; Hatakeyama, K.; Quitain, A.T.; Kida, T. Carbon-based potentiometric hydrogen sensor using a proton conducting graphene oxide membrane coupled with a WO_3 sensing electrode. *Sens. Actuators B Chem.* **2020**, *323*, 128678. [CrossRef]
69. Rasch, F.; Postica, V.; Schütt, F.; Mishra, Y.K.; Nia, A.S.; Lohe, M.R.; Feng, X.; Adelung, R.; Lupan, O. Highly selective and ultra-low power consumption metal oxide based hydrogen gas sensor employing graphene oxide as molecular sieve. *Sens. Actuators B Chem.* **2020**, *320*, 128363. [CrossRef]
70. Wang, S.; Yan, H.; Zheng, H.; He, Y.; Guo, X.; Li, S.; Yang, C. Fast response humidity sensor based on chitosan/graphene oxide/tin dioxide composite. *Sens. Actuators B Chem.* **2023**, *392*, 134070. [CrossRef]
71. Zeng, S.; Pan, Q.; Huang, Z.; Gu, C.; Wang, T.; Xu, J.; Yan, Z.; Zhao, F.; Li, P.; Tu, Y.; et al. Ultrafast response of self-powered humidity sensor of flexible graphene oxide film. *Mater. Des.* **2023**, *226*, 111683. [CrossRef]
72. Morsy, M.; Ibrahim, M.; Yuan, Z.; Meng, F. Graphene Foam Decorated With ZnO as a Humidity Sensor. *IEEE Sens. J.* **2020**, *20*, 1721–1729. [CrossRef]
73. Xuan, W.; He, M.; Meng, N.; He, X.; Wang, W.; Chen, J.; Shi, T.; Hasan, T.; Xu, Z.; Xu, Y.; et al. Fast Response and High Sensitivity ZnO/glass Surface Acoustic Wave Humidity Sensors Using Graphene Oxide Sensing Layer. *Sci. Rep.* **2014**, *4*, 7206. [CrossRef] [PubMed]
74. Li, N.; Chen, X.; Chen, X.; Ding, X.; Zhao, X. Ultrahigh humidity sensitivity of graphene oxide combined with Ag nanoparticles. *Rsc Adv.* **2017**, *7*, 45988–45996. [CrossRef]
75. Shen, Q.-Q.; Zhang, C.-Z.; Bai, Y.; Ni, M.-R. Synthesizing N–[4-morpholinecarboximidamidoyl]carboximidamidoylated graphene oxide for fabricating high-sensitive humidity sensors. *Diam. Relat. Mater.* **2022**, *126*, 109053. [CrossRef]
76. Song, H.; Li, X.; Cui, P.; Guo, S.; Liu, W.; Wang, X. Sensitivity investigation for the dependence of monolayer and stacking graphene NH_3 gas sensor. *Diam. Relat. Mater.* **2017**, *73*, 56–61. [CrossRef]
77. Su, P.-G.; Liao, Z.-H. Fabrication of a flexible single-yarn NH_3 gas sensor by layer-by-layer self-assembly of graphene oxide. *Mater. Chem. Phys.* **2019**, *224*, 349–356. [CrossRef]
78. Park, M.-S.; Kim, K.H.; Kim, M.-J.; Lee, Y.-S. NH_3 gas sensing properties of a gas sensor based on fluorinated graphene oxide. *Colloids Surf. A Physicochem. Eng. Asp.* **2016**, *490*, 104–109. [CrossRef]
79. Li, Q.; Sun, M.; Jiang, C.; Song, S.; Li, T.; Xu, M.; Chen, W.; Peng, H. Phosphorus doping of graphene for conductometric room temperature ammonia sensing. *Sens. Actuators B Chem.* **2023**, *379*, 133234. [CrossRef]
80. Huang, X.; Hu, N.; Zhang, L.; Wei, L.; Wei, H.; Zhang, Y. The NH_3 sensing properties of gas sensors based on aniline reduced graphene oxide. *Synth. Met.* **2013**, *185–186*, 25–30. [CrossRef]
81. Raza, A.; Abid, R.; Murtaza, I.; Fan, T. Room temperature NH_3 gas sensor based on PMMA/RGO/ZnO nanocomposite films fabricated by in-situ solution polymerization. *Ceram. Int.* **2023**, *49*, 27050–27059. [CrossRef]
82. Achary, L.S.K.; Kumar, A.; Barik, B.; Nayak, P.S.; Tripathy, N.; Kar, J.P.; Dash, P. Reduced graphene oxide-$CuFe_2O_4$ nanocomposite: A highly sensitive room temperature NH_3 gas sensor. *Sens. Actuators B Chem.* **2018**, *272*, 100–109. [CrossRef]

83. Matatagui, D.; López-Sánchez, J.; Peña, A.; Serrano, A.; del Campo, A.; de la Fuente, O.R.; Carmona, N.; Navarro, E.; Marín, P.; del Carmen Horrillo, M. Ultrasensitive NO$_2$ gas sensor with insignificant NH$_3$-interference based on a few-layered mesoporous graphene. *Sens. Actuators B Chem.* **2021**, *335*, 129657. [CrossRef]
84. Niu, F.; Shao, Z.-W.; Gao, H.; Tao, L.-M.; Ding, Y. Si-doped graphene nanosheets for NOx gas sensing. *Sens. Actuators B Chem.* **2021**, *328*, 129005. [CrossRef]
85. Ye, X.; Qi, M.; Qiang, Y.; Chen, M.; Zheng, X.; Gu, M.; Zhao, X.; Yang, Y.; He, C.; Zhang, J. Laser-ablated violet phosphorus/graphene heterojunction as ultrasensitive ppb-level room-temperature NO sensor. *Chin. Chem. Lett.* **2023**, *34*, 108199. [CrossRef]
86. Duy, L.T.; Noh, Y.G.; Seo, H. Improving graphene gas sensors via a synergistic effect of top nanocatalysts and bottom cellulose assembled using a modified filtration technique. *Sens. Actuators B Chem.* **2021**, *334*, 129676. [CrossRef]
87. Liu, S.; Zhou, L.; Yao, L.; Chai, L.; Li, L.; Zhang, G.; Kankan; Shi, K. One-pot reflux method synthesis of cobalt hydroxide nanoflake-reduced graphene oxide hybrid and their NOx gas sensors at room temperature. *J. Alloys Compd.* **2014**, *612*, 126–133. [CrossRef]
88. Zhang, C.; Zhang, S.; Zhang, D.; Yang, Y.; Zhao, J.; Yu, H.; Wang, T.; Wang, T.; Dong, X. Conductometric room temperature NOx sensor based on metal-organic framework-derived Fe$_2$O$_3$/Co$_3$O$_4$ nanocomposite. *Sens. Actuators B Chem.* **2023**, *390*, 133894. [CrossRef]
89. Pungjunun, K.; Chaiyo, S.; Praphairaksit, N.; Siangproh, W.; Ortner, A.; Kalcher, K.; Chailapakul, O.; Mehmeti, E. Electrochemical detection of NOx gas based on disposable paper-based analytical device using a copper nanoparticles-modified screen-printed graphene electrode. *Biosens. Bioelectron.* **2019**, *143*, 111606. [CrossRef]
90. Ugale, A.D.; Umarji, G.G.; Jung, S.H.; Deshpande, N.G.; Lee, W.; Cho, H.K.; Yoo, J.B. ZnO decorated flexible and strong graphene fibers for sensing NO$_2$ and H$_2$S at room temperature. *Sens. Actuators B Chem.* **2020**, *308*, 127690. [CrossRef]
91. Song, Z.; Wei, Z.; Wang, B.; Luo, Z.; Xu, S.; Zhang, W.; Yu, H.; Li, M.; Huang, Z.; Zang, J.; et al. Sensitive Room-Temperature H$_2$S Gas Sensors Employing SnO$_2$ Quantum Wire/Reduced Graphene Oxide Nanocomposites. *Chem. Mater.* **2016**, *28*, 1205–1212. [CrossRef]
92. Liu, L.; Yang, M.; Gao, S.; Zhang, X.; Cheng, X.; Xu, Y.; Zhao, H.; Huo, L.; Major, Z. Co$_3$O$_4$ Hollow Nanosphere-Decorated Graphene Sheets for H$_2$S Sensing near Room Temperature. *ACS Appl. Nano Mater.* **2019**, *2*, 5409–5419. [CrossRef]
93. Kumar, R.; Kaur, A. Chemiresistive gas sensors based on thermally reduced graphene oxide for sensing sulphur dioxide at room temperature. *Diam. Relat. Mater.* **2020**, *109*, 108039. [CrossRef]
94. Van Cat, V.; Dinh, N.X.; Ngoc Phan, V.; Le, A.T.; Nam, M.H.; Dinh Lam, V.; Dang, T.V.; Quy, N.V. Realization of graphene oxide nanosheets as a potential mass-type gas sensor for detecting NO$_2$, SO$_2$, CO, and NH$_3$. *Mater. Today Commun.* **2020**, *25*, 101682. [CrossRef]
95. Kumar, R.; Avasthi, D.; Kaur, A. Fabrication of chemiresistive gas sensors based on multistep reduced graphene oxide for low parts per million monitoring of sulfur dioxide at room temperature. *Sens. Actuators B Chem.* **2017**, *242*, 461–468. [CrossRef]
96. Wei, Y.; Zhang, P.; Soomro, R.A.; Zhu, Q.; Xu, B. Advances in the synthesis of 2D MXenes. *Adv. Mater.* **2021**, *33*, 2103148. [CrossRef] [PubMed]
97. Smith, A.T.; LaChance, A.M.; Zeng, S.; Liu, B.; Sun, L. Synthesis, properties, and applications of graphene oxide/reduced graphene oxide and their nanocomposites. *Nano Mater. Sci.* **2019**, *1*, 31–47. [CrossRef]
98. Alam, S.N.; Sharma, N.; Kumar, L. Synthesis of graphene oxide (GO) by modified hummers method and its thermal reduction to obtain reduced graphene oxide (rGO). *Graphene* **2017**, *6*, 1–18. [CrossRef]
99. Razaq, A.; Bibi, F.; Zheng, X.; Papadakis, R.; Jafri, S.H.M.; Li, H. Review on graphene-, graphene oxide-, reduced graphene oxide-based flexible composites: From fabrication to applications. *Materials* **2022**, *15*, 1012. [CrossRef]
100. Liu, J.; Tang, J.; Gooding, J.J. Strategies for chemical modification of graphene and applications of chemically modified graphene. *J. Mater. Chem.* **2012**, *22*, 12435–12452. [CrossRef]
101. Todorovic, A. Gases and vapours. In *Principles of Occupational Health and Hygiene*; Routledge: London, UK, 2020; pp. 242–282.
102. del Castillo, R.M.; Calles, A.G.; Espejel-Morales, R.; Hernández-Coronado, H. Adsorption of CO$_2$ on graphene surface modified with defects. *Comput. Condens. Matter* **2018**, *16*, e00315. [CrossRef]
103. Deji; Kaur, N.; Choudhary, B.C.; Sharma, R.K. Carbon-dioxide gas sensor using co-doped graphene nanoribbon: A first principle DFT study. *Mater. Today Proc.* **2021**, *45*, 5023–5028. [CrossRef]
104. Elgammal, K.; Hugosson, H.W.; Smith, A.D.; Råsander, M.; Bergqvist, L.; Delin, A. Density functional calculations of graphene-based humidity and carbon dioxide sensors: Effect of silica and sapphire substrates. *Surf. Sci.* **2017**, *663*, 23–30. [CrossRef]
105. Dindorkar, S.S.; Yadav, A. Comparative study on adsorption behaviour of the monolayer graphene, boron nitride and silicon carbide hetero-sheets towards carbon monoxide: Insights from first-principle studies. *Comput. Theor. Chem.* **2022**, *1211*, 113676. [CrossRef]
106. Deji, V.; Akarsh Kaur, N.; Choudhary, B.C.; Sharma, R.K. Adsorption chemistry of co-doped graphene nanoribbon and its derivatives towards carbon based gases for gas sensing applications: Quantum DFT investigation. *Mater. Sci. Semicond. Process.* **2022**, *146*, 106670. [CrossRef]
107. Deji, R.; Verma, A.; Kaur, N.; Choudhary, B.C.; Sharma, R.K. Density functional theory study of carbon monoxide adsorption on transition metal doped armchair graphene nanoribbon. *Mater. Today Proc.* **2022**, *54*, 771–776. [CrossRef]

108. Wang, C.; Wang, Y.; Yang, Z.; Hu, N. Review of recent progress on graphene-based composite gas sensors. *Ceram. Int.* **2021**, *47*, 16367–16384. [CrossRef]
109. Lachawiec, A.J.; Qi, G.; Yang, R.T. Hydrogen Storage in Nanostructured Carbons by Spillover: Bridge-Building Enhancement. *Langmuir* **2005**, *21*, 11418–11424. [CrossRef]
110. Kumar, N.; Jasani, J.; Sonvane, Y.; Korvink, J.G.; Sharma, A.; Sharma, B. Unfolding the hydrogen gas sensing mechanism across 2D Pnictogen/graphene heterostructure sensors. *Sens. Actuators B Chem.* **2024**, *399*, 134807. [CrossRef]
111. Frediani, M.; Oberhauser, W.; Rosi, L.; Bartoli, M.; Passaglia, E.; Capozzoli, L. Palladium nanoparticles supported onto stereocomplexed poly (lactic acid)-poly (ε-caprolactone) copolymers for selective partial hydrogenation of phenylacetylene. *Rend. Lincei* **2017**, *28*, 51–58. [CrossRef]
112. Mattia Bartoli, L.R.; Petrucci, G.; Armelao, L.; Oberhauser, W.; Frediani, M.; Piccoloe, O.; Rathod, V.D.; Paganelli, S. An easily recoverable and recyclable homogeneous polyester-based Pd catalytic system for the hydrogenation of α,β-unsaturated carbonyl compounds. *Catal. Commun.* **2015**, *69*, 228–233. [CrossRef]
113. Oberhauser, W.; Bartoli, M.; Petrucci, G.; Bandelli, D.; Frediani, M.; Capozzoli, L.; Cepek, C.; Bhardwaj, S.; Rosi, L. Nitrile hydration to amide in water: Palladium-based nanoparticles vs molecular catalyst. *J. Mol. Catal. A Chem.* **2015**, *410*, 26–33. [CrossRef]
114. Oberhauser, W.; Evangelisti, C.; Jumde, R.; Petrucci, G.; Bartoli, M.; Frediani, M.; Mannini, M.; Capozzoli, L.; Passaglia, E.; Rosi, L. Palladium-nanoparticles on end-functionalized poly(lactic acid)-based stereocomplexes for the chemoselective cinnamaldehyde hydrogenation: Effect of the end-group. *J. Catal.* **2015**, *330*, 187–196. [CrossRef]
115. Oberhauser, W.; Evangelisti, C.; Tiozzo, C.; Bartoli, M.; Frediani, M.; Passaglia, E.; Rosi, L. Platinum nanoparticles onto pegylated poly (lactic acid) stereocomplex for highly selective hydrogenation of aromatic nitrocompounds to anilines. *Appl. Catal. A Gen.* **2017**, *537*, 50–58. [CrossRef]
116. Petrucci, G.; Oberhauser, W.; Bartoli, M.; Giachi, G.; Frediani, M.; Passaglia, E.; Capozzoli, L.; Rosi, L. Pd-nanoparticles supported onto functionalized poly(lactic acid)-based stereocomplexes for partial alkyne hydrogenation. *Appl. Catal. A Gen.* **2014**, *469*, 132–138. [CrossRef]
117. Bartoli, M.; Rosi, L.; Mini, B.; Petrucci, G.; Passaglia, E.; Frediani, M. Catalytic Performances of Platinum Containing PLLA Macrocomplex in the Hydrogenation of α, β-Unsaturated Carbonyl Compounds. *Appl. Sci.* **2019**, *9*, 3243. [CrossRef]
118. Wang, L.; Li, W.; Cai, Y.; Pan, P.; Li, J.; Bai, G.; Xu, J. Characterization of Pt-or Pd-doped graphene based on density functional theory for H2 gas sensor. *Mater. Res. Express* **2019**, *6*, 095603. [CrossRef]
119. Kishnani, V.; Yadav, A.; Mondal, K.; Gupta, A. Palladium-Functionalized Graphene for Hydrogen Sensing Performance: Theoretical Studies. *Energies* **2021**, *14*, 5738. [CrossRef]
120. Lange, U.; Hirsch, T.; Mirsky, V.M.; Wolfbeis, O.S. Hydrogen sensor based on a graphene—Palladium nanocomposite. *Electrochim. Acta* **2011**, *56*, 3707–3712. [CrossRef]
121. Yu, L.; Gao, W.; R Shamshiri, R.; Tao, S.; Ren, Y.; Zhang, Y.; Su, G. Review of research progress on soil moisture sensor technology. *Int. J. Agric. Biol. Eng.* **2021**, *14*, 32–42. [CrossRef]
122. Ye, X.; Qi, M.; Yang, H.; Mediko, F.S.; Qiang, H.; Yang, Y.; He, C. Selective sensing and mechanism of patterned graphene-based sensors: Experiments and DFT calculations. *Chem. Eng. Sci.* **2022**, *247*, 117017. [CrossRef]
123. Agmon, N. The grotthuss mechanism. *Chem. Phys. Lett.* **1995**, *244*, 456–462. [CrossRef]
124. Seeneevassen, S.; Leong, A.; Kashan, M.A.M.; Swamy, V.; Ramakrishnan, N. Effect of effluent gas composition on characteristics of graphene oxide film based relative humidity sensor. *Measurement* **2022**, *195*, 111156. [CrossRef]
125. Huang, Y.; Zeng, Z.; Liang, T.; Li, J.; Liao, Z.; Li, J.; Yang, T. An encapsulation strategy of graphene humidity sensor for enhanced anti-interference ability. *Sens. Actuators B Chem.* **2023**, *396*, 134517. [CrossRef]
126. Yarandi, M.S.; Mahdinia, M.; Barazandeh, J.; Soltanzadeh, A. Evaluation of the toxic effects of ammonia dispersion: Consequence analysis of ammonia leakage in an industrial slaughterhouse. *Med. Gas Res.* **2021**, *11*, 24. [PubMed]
127. Ojha, M.; Dhiman, A. Problem, failure and safety analysis of ammonia plant: A review. *Int. Rev. Chem. Eng.* **2010**, *2*, 631–646.
128. Chen, G.; Gan, L.; Xiong, H.; Zhang, H. Density Functional Theory Study of B, N, and Si Doped Penta-Graphene as the Potential Gas Sensors for NH_3 Detection. *Membranes* **2022**, *12*, 77. [CrossRef]
129. Choudhari, U.; Jagtap, S. A panoramic view of NOx and NH_3 gas sensors. *Nano-Struct. Nano-Objects* **2023**, *35*, 100995. [CrossRef]
130. Xiao, H.-M.; Hou, Y.-C.; Guo, Y.-R.; Pan, Q.-J. The coupling of graphene, graphitic carbon nitride and cellulose to fabricate zinc oxide-based sensors and their enhanced activity towards air pollutant nitrogen dioxide. *Chemosphere* **2023**, *324*, 138325. [CrossRef]
131. Ayesh, A.I. The effect of ZrOx modification of graphene nanoribbon on its adsorption for NOx: A DFT investigation. *Mater. Chem. Phys.* **2022**, *291*, 126693. [CrossRef]
132. Dwivedi, G.; Deshwal, M.; Kishor Johar, A. ZnO based NOx gas and VOC detection sensor fabrication techniques and materials. *Mater. Today Proc.* **2023**. [CrossRef]
133. Wang, M.; Chen, D.; Jia, P. Adsorption and sensing performances of air decomposition components (CO, NOx) on Cr modified graphene surface. *Inorg. Chem. Commun.* **2023**, *157*, 111447. [CrossRef]
134. Reshak, A.; Auluck, S. Adsorbing H_2S onto a single graphene sheet: A possible gas sensor. *J. Appl. Phys.* **2014**, *116*, 103702. [CrossRef]
135. Tan, G.-L.; Tang, D.; Wang, X.-M.; Yin, X.-T. Overview of the Recent Advancements in Graphene-Based H_2S Sensors. *ACS Appl. Nano Mater.* **2022**, *5*, 12300–12319. [CrossRef]

136. Qiao, Y.; Li, X.; Hirtz, T.; Deng, G.; Wei, Y.; Li, M.; Ji, S.; Wu, Q.; Jian, J.; Wu, F. Graphene-based wearable sensors. *Nanoscale* **2019**, *11*, 18923–18945. [CrossRef] [PubMed]
137. Park, J.; Kim, J.; Kim, K.; Kim, S.-Y.; Cheong, W.H.; Park, K.; Song, J.H.; Namgoong, G.; Kim, J.J.; Heo, J.; et al. Wearable, wireless gas sensors using highly stretchable and transparent structures of nanowires and graphene. *Nanoscale* **2016**, *8*, 10591–10597.
138. Singh, E.; Meyyappan, M.; Nalwa, H.S. Flexible graphene-based wearable gas and chemical sensors. *ACS Appl. Mater. Interfaces* **2017**, *9*, 34544–34586. [CrossRef] [PubMed]
139. Xu, H.; Xiang, J.X.; Lu, Y.F.; Zhang, M.K.; Li, J.J.; Gao, B.B.; Zhao, Y.J.; Gu, Z.Z. Multifunctional wearable sensing devices based on functionalized graphene films for simultaneous monitoring of physiological signals and volatile organic compound biomarkers. *ACS Appl. Mater. Interfaces* **2018**, *10*, 11785–11793. [CrossRef]
140. Peng, Z.; Tao, L.-Q.; Zou, S.; Zhu, C.; Wang, G.; Sun, H.; Ren, T.-L. A Multi-functional NO_2 gas monitor and Self-Alarm based on Laser-Induced graphene. *Chem. Eng. J.* **2022**, *428*, 131079. [CrossRef]
141. Li, Q.; Wu, T.; Zhao, W.; Li, Y.; Ji, J.; Wang, G. 3D printing stretchable core-shell laser scribed graphene conductive network for self-powered wearable devices. *Compos. Part B Eng.* **2022**, *240*, 110000. [CrossRef]
142. Sun, B.; McCay, R.N.; Goswami, S.; Xu, Y.; Zhang, C.; Ling, Y.; Lin, J.; Yan, Z. Gas-Permeable, Multifunctional On-Skin Electronics Based on Laser-Induced Porous Graphene and Sugar-Templated Elastomer Sponges. *Adv. Mater.* **2018**, *30*, 1804327. [CrossRef]

Disclaimer/Publisher's Note: The statements, opinions and data contained in all publications are solely those of the individual author(s) and contributor(s) and not of MDPI and/or the editor(s). MDPI and/or the editor(s) disclaim responsibility for any injury to people or property resulting from any ideas, methods, instructions or products referred to in the content.

Article

Second-Order Nonlinearity of Graphene Quantum Dots Measured by Hyper-Rayleigh Scattering

Manoel L. Silva-Neto [1,†], Renato Barbosa-Silva [2], Georges Boudebs [3,*] and Cid B. de Araújo [2]

1 Programa de Pós-Graduação em Ciência de Materiais, Universidade Federal de Pernambuco, Recife 50670-901, PE, Brazil; manoel.neto@utoronto.ca
2 Departamento de Física, Universidade Federal de Pernambuco, Recife 50670-901, PE, Brazil; cid.araujo@ufpe.br (C.B.d.A.)
3 Univ Angers, LPHIA, SFR MATRIX, F 49000 Angers, France
* Correspondence: georges.boudebs@univ-angers.fr
† Current address: Department of Chemistry, University of Toronto, 80 St. George Street, Toronto, ON M5S3H6, Canada.

Abstract: The first hyperpolarizability of graphene quantum dots (GQDs) suspended in water was determined using the hyper-Rayleigh scattering (HRS) technique. To the best of our knowledge, this is the first application of the HRS technique to characterize GQDs. Two commercial GQDs (Acqua-Cyan and Acqua-Green) with different compositions were studied. The HRS experiments were performed with an excitation laser at 1064 nm. The measured hyperpolarizabilities were $(1.0 \pm 0.1) \times 10^{-27}$ esu and $(0.9 \pm 0.1) \times 10^{-27}$ esu for Acqua-Cyan and Acqua-Green, respectively. The results were used to estimate the hyperpolarizability per nanosheet obtained by assuming that each GQD has five nanosheets with 0.3 nm thickness. The two-level model, used to calculate the static hyperpolarizability per nanosheet, provides values of $(2.4 \pm 0.1) \times 10^{-28}$ esu (Acqua-Cyan) and $(0.5 \pm 0.1) \times 10^{-28}$ esu (Aqua-Green). The origin of the nonlinearity is discussed on the basis of polarized resolved HRS experiments, and electric quadrupolar behavior with a strong dependence on surface effects. The nontoxic characteristics and order of magnitude indicate that these GQDs may be useful for biological microscopy imaging.

Keywords: hyper-Rayleigh scattering; graphene quantum dots; second-harmonic generation

Citation: Silva-Neto, M.L.; Barbosa-Silva, R.; Boudebs, G.; de Araújo, C.B. Second-Order Nonlinearity of Graphene Quantum Dots Measured by Hyper-Rayleigh Scattering. Materials 2023, 16, 7376. https://doi.org/10.3390/ma16237376

Academic Editors: Victoria Samanidou and Eleni Deliyanni

Received: 20 October 2023
Revised: 20 November 2023
Accepted: 24 November 2023
Published: 27 November 2023

Copyright: © 2023 by the authors. Licensee MDPI, Basel, Switzerland. This article is an open access article distributed under the terms and conditions of the Creative Commons Attribution (CC BY) license (https://creativecommons.org/licenses/by/4.0/).

1. Introduction

The hyper-Rayleigh scattering (HRS) technique has been employed for several years to investigate the nonlinear optical properties of molecules in liquid suspensions as well as colloidal metal particles in aqueous solutions, nanocrystals in liquid suspensions, and various other materials. It entails directing a laser beam onto a sample and detecting the scattered light, that has double the frequency of the incident light. The intensity of the scattered light is contingent on the first hyperpolarizability of the sample molecular property (individual particles in our context), a measure of their ability to alter the polarization of light. In principle, the HRS technique provides a direct assessment of the magnitude of the first hyperpolarizability of molecules and nanoparticles that describes the second-order nonlinear optical (NLO) response. Substantial NLO activity is a crucial prerequisite for emerging technological photonic applications such as photonics and all-optical switching and optical parametric oscillators, which have intensified the quest for NLO-active materials. For example, previous research has underscored the remarkable NLO response in solutions of metallic nanometer-sized particles in solutions. The origin of this response lies in the one- or two-photon resonance of the incident laser light used in the experiments, coupled with the particles' strong surface plasmon absorption. Graphene quantum dots (GQDs) are nanometer-size graphene segments that are small enough to exhibit exciton confinement and quantum size effects. Unlike graphene, they present a non-zero energy bandgap that is responsible for their characteristic electrical and optical properties.

Generally, the optical absorption spectra of GQDs present a strong peak in the ultraviolet ($\approx$ 230 nm) due to the $\pi \to \pi^*$ excitation of the π bonds of aromatic C=C and a weaker peak at $\approx$ 300 nm due to $n \to \pi^*$ transitions of C=O bonds [1,2]. The introduction of functional groups into the GQDs may lead to new absorption features and may change the GQDs' optical properties [3,4]. Moreover, since they present molecule-like structures the bandgap can be tuned by changing the GQD size and by surface chemistry. Therefore, the optical properties of GQDs are dependent on the preparation method as well as the functional groups added to the graphene segments [3–8].

When compared to heavy-metal-based semiconductor quantum dots that cause concern for in vivo bioimaging, GQDs, due to their low toxicity [9,10] and strong two-photon induced fluorescence, have been exploited for biological applications by many authors [3,11,12]. Likewise, there are reported applications of GQDs for energy conversion, environmental monitoring, and electronic sensors for humidity detection, among other applications [3,4,7,13,14]. Recently, Qi et al. reported second harmonic generation (SHG) in boron doped-graphene quantum dots (B-GQD) and their application in stem cell imaging and tracking wound healing. These applications were achieved because the doping promoted the break of symmetry in the B-GQD allowing the SHG signal [15].

Although optical absorption, linear luminescence, second harmonic generation, and the third-order nonlinearity of GQDs have been studied [5–8,16–18], the second-order nonlinearity, associated with the first hyperpolarizability, $\beta(\omega)$, did not receive much attention. It is true that the experiments to be performed are not very easy to set up, such as measuring nonlinear refraction in a 4f setup (see for example [19]) to characterize the third-order NL response. Computational results were reported in [20–22] for some GQDs but no experimental result is available. Here, we want to estimate the order of magnitude of these coefficients knowing that the manufacturing methods relative to new materials, sometimes different from one lab to another, could influence slightly the result. The absorption spectra of certain DIY items, assumed to be identical, exhibit partial mismatches owing to the presence of significant residues (impurities) within the final solution. Hence, it proves beneficial to examine readily accessible, commercially available samples. Preparations employing hybrid manufacturing methods may yield slightly divergent results but should not alter the core conclusion in this context.

In the present work, the hyper-Rayleigh scattering (HRS) technique [23,24] was employed for the first time to determine the effective first hyperpolarizability, $\langle \beta(2\omega) \rangle_{eff}$, of GQDs suspended in water. The HRS is an incoherent nonlinear process in which two photons with frequency ω are instantaneously combined to generate a new photon with frequency 2ω due to the interaction with a material. Unlike the two-photon induced luminescence process, the generation of second harmonic in the HRS process does not require on-resonance excitation of the medium to high energy levels. The HRS technique has been routinely used to study molecules [23–27], metal nanoparticles in liquid suspensions [28–30], and dielectric nanocrystals [31–35]. The first hyperpolarizabilities of molecules and nanoparticles can be determined as well as the origin of the incoherent second harmonic generation (SHG) signal detected.

Studying the second-order nonlinearity of graphene quantum dots is not only essential for advancing our fundamental understanding of these nanomaterials but also for unlocking their potential in a wide range of practical applications, from optoelectronics to material science and beyond. The origin of the second-order nonlinear effect arising from nanoparticles (NPs) with crystalline structure presenting inversion symmetry was discussed in [36–39]. Two important cases to consider are: NPs with noncentrosymmetric shape and NPs with centrosymmetric shape [39]. In the former, SHG is due to the induced electric dipole in the NP volume and in the interfaces. In the latter, the induced electric dipole does not occur because of the inversion symmetry in the crystalline structure of the NPs. Then, besides the contributions from the interfaces, an important mechanism of SHG is related to the field's gradients occurring within the NP bulk [38,39]. However, it is difficult to synthesize NPs with perfect shape, and therefore, in real NPs, there is competition between the two main cases above mentioned.

In the present paper, we report values for the first hyperpolarizabilities of two different GQDs [40] investigated by measuring the intensity of the incoherent SHG signal upon excitation using a laser off-resonance with absorption transitions of the GQDs. Values of $\langle \beta(2\omega) \rangle_{eff}$ were determined by applying the HRS technique used for the calibration of para-nitroaniline (p-NA), which is the usual reference standard for HRS measurements. To determine the static hyperpolarizability per nanosheet, $\beta(0)^{NS}$, the classical two-level model [35,41] was applied in combination with the results for $\langle \beta(2\omega) \rangle_{eff}$ and considering that each GQD has five nanosheets with a thickness of 0.3 nm each. Studying the non-linear optical properties of these materials can provide insights into their structure and functionality involving the interaction of light with nanoparticles at a nanoscale level.

2. Experimental Details

The experiments were performed with commercially available GQDs (Acqua-Green and Acqua-Cyan) purchased from STREM Chemical Incorporation (Kehl, Germany) [40]. Typically, these quantum dots, composed of graphene nanosheets, have diameters of about (5 ± 1) nm and average height of (1.5 ± 0.5) nm, corresponding to five graphene sheets. The GQD concentration in the present experiments was 3.4×10^{16} particles/cm^3 for both samples that were suspended in water. Experiments were also performed with more diluted samples to verify their behavior with the laser intensity. The concentration of nanosheets per cm^3 was estimated considering the data sheet provided by the manufacturer. The absorbance spectra of the samples were measured with a commercial spectrophotometer from 200 to 1500 nm. The photoluminescence (PL) spectra and their temporal evolution were measured using a fast photodetector with a time resolution of 1 ns.

Figure 1 shows the setup used for the HRS experiments. The excitation source was a Q-switched Nd:YAG laser (1064 nm, 6 ns, 10 Hz). A bandpass filter (F1) was used to avoid scattered light from the fundamental beam. A half-wave plate ($\lambda/2$), a polarizer (P), a beam splitter (BS), and a reference photodetector (RP) were employed to control the incident laser intensity on the sample. A 5 cm focal length lens (L1) was used to focus the laser beam on the sample that was contained in a 1 cm long quartz cuvette. The HRS signal was collected perpendicularly to the laser beam direction by using two lenses (focal length: 5 cm) to focus the HRS signal on the photomultiplier (PMT) that was coupled to a spectrometer. This setup allowed for the analysis of the spectral content of the scattered light. Experiments were also performed using an interference filter (F2) centered at 532 nm with full width at half maximum (FWHM) of 8 nm. The photomultiplier was connected to an oscilloscope and a computer for data collection. For temporal analysis of the scattered light, the SHG signal was focused on a fast photodetector (response time: 1 ns) instead of the spectrometer/photomultiplier. For the polarization dependence measurements, another half-wave ($\lambda/2$) plate and another polarizer (AP) were inserted in the experimental setup shown in Figure 1 and the measurements were performed following the procedure of [28–30]. The half-wave plate used for polarization dependence measurements was inserted between the beam splitter (BS) and lenses L1. It is important to mention that the half-wave was coupled to a stepper motor controlled by a Labview-based program. For the measurements, the stepper motor is rotated in increments of 0–180 degrees. The polarizer (AP) was inserted between F2 and the spectrometer. During the measurement, the second polarizer is held in the vertical (V) or horizontal (H) polarization. Para-nitroaniline (p-NA) dissolved in methanol was used as a reference standard [23].

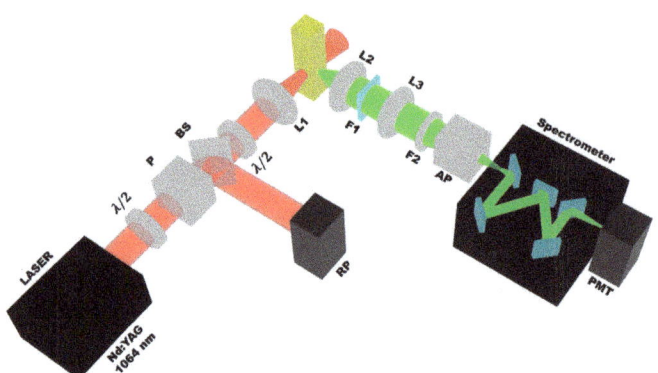

Figure 1. Experimental setup used for the hyper-Rayleigh scattering experiments. P is a Polarizer, BS is a Beam Splitter, RP is the Reference Photodetector, L1 is a focusing lens, L2 and L3 are collecting lenses, F2 is an interference filter, AP is an Analyzer Polarizer mounted in a rotation stage in a way that allows to set up the polarizer axis as Vertical (V) or Horizontal (H). PMT is a Photomultiplier Tube.

3. Results and Discussion

Figure 2a shows the absorbance spectra of both GQDs samples. The absorption bands in the ultraviolet are associated with the $\pi \to \pi^*$ and $n \to \pi^*$ transitions [1,2]. The arrows in Figure 2a indicate the laser and the HRS wavelengths. Notice that, for both samples, the laser and its second harmonic are off-resonance with the absorption bands; therefore, we do not expect a PL signal superimposed on the HRS signal. Figure 2b shows the normalized PL spectra obtained by excitation at 386 nm for the Acqua-Cyan GQDs and 486 nm for the Acqua-Green GQDs using a commercial spectrofluorometer (Fluoromax HORIBA, Kyoto, Japan). The temporal decay of the PL signal with a maximum at 550 nm (Acqua-Green) and 480 nm (Acqua-Cyan), excited at 850 nm (via two-photon absorption), was fit by a single exponential with a decay time of 26 ns (Acqua-Green) and 19 ns (Acqua-Cyan).

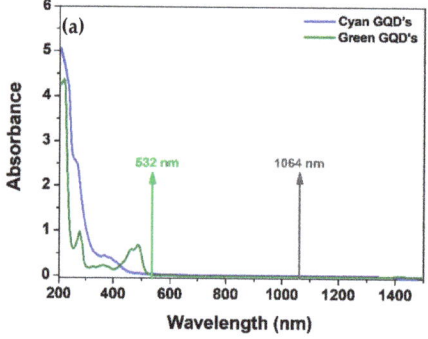
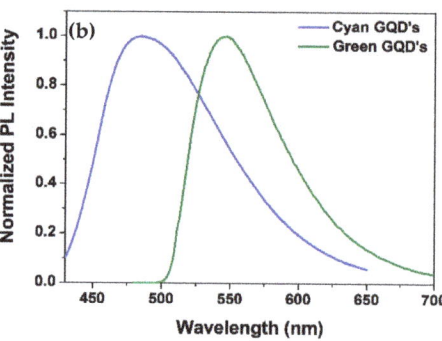

Figure 2. (**a**) Optical absorbance spectra; (**b**) One-photon induced photoluminescence spectra (excitation at 386 nm and 485 nm for Acqua-Cyan and Acqua-Green GQDs, respectively). The normalization factors for luminescence are 9.5×10^5 for Acqua-Cyan GQDs and 1.7×10^6 Acqua-Green GQDs. Samples concentrations: 3.4×10^{16} particles/cm^3.

Figure 3 shows the HRS spectra and the temporal behavior of the HRS signal (obtained for a single laser shot). The solid red lines in Figure 3 are the Gaussian fit to the experimental data and the solid black lines on the inset of Figure 3b show the temporal evolution of the excitation laser. The duration for both GQDs was determined to be 4.6 ns from a Gaussian fit. Notice in Figure 3a that the spectra are centered at 532 nm, the second harmonic of the incident laser beam, with the linewidth limited by the spectrometer resolution. The temporal evolution of the HRS signal exhibited in the inset of Figure 3b follows the laser pulse corroborating the above statement that no long-lived luminescence overlaps with the HRS signal. The HRS is a parametric phenomenon, and the strong correlation between the temporal behavior of the signal and the excitation source supports this assertion. These measurements were performed with a laser peak intensity of 12 GW/cm^2, and the minimum intensity that allowed the detection of the HRS signal was 8.0 GW/cm^2.

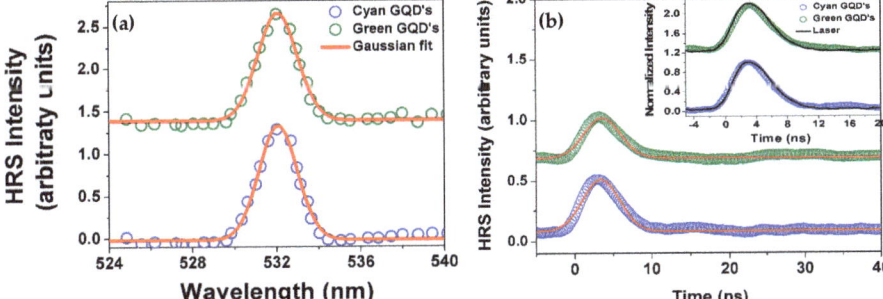

Figure 3. (**a**) Spectra of the incoherent second harmonic scattered light for excitation at 1064 nm. The continuous lines are Gaussian fits. (**b**) The green and blue circles show the temporal HRS data for both samples (single laser shot). The curves are shifted in the vertical direction to allow better visualization. The solid red lines are Gaussian fits to the data. The inset in (**b**) shows a comparison between the HRS signal and the excitation source. The green and blue circles in the inset show the temporal data for both samples and the black line illustrates the laser pulse temporal profile. The temporal response of the photodetector is 1 ns.

The HRS signal is due to the nonlinear scattering by the randomly oriented GQDs and water molecules. The scattered intensity, at the second harmonic frequency of the incident laser beam, is described by $I(2\omega) = g \sum_c N_c \langle \beta_c^2(2\omega) \rangle I^2(\omega)$, where $I(\omega)$ is the laser intensity, $\langle \beta_c^2(2\omega) \rangle = \langle \beta_c^2(2\omega; \omega; \omega) \rangle$ is the orientational average of the first hyperpolarizability and the sub-index represents the constituents of the suspension. The symbol $\langle \rangle$ indicates orientational average. The factor g depends on the scattering geometry and contains information on the transformation of coordinates from the GQD to the laboratory reference system [31–35].

Figure 4a shows plots of the HRS signal, $S(2\omega) \propto I(2\omega)$, versus the incident laser intensity where a quadratic dependence with $I(\omega)$ was verified. Figure 4b shows the linear dependence of $S(2\omega)$ versus the GQDs concentration, N_{GQD}, which confirms the contribution of isolated GQDs to the HRS signal. Moreover, we recall that aggregation of GQDs would redshift the linear absorption spectrum, a behavior that was not observed for the various concentrations used.

Since the samples consist of GQDs with different dimensions, the experiments provided effective values for the GQD hyperpolarizability, $\langle \beta^2(2\omega) \rangle_{eff}$. The external reference method [42], using p-NA dissolved in methanol as the reference standard, was applied to measure $\langle \beta(2\omega) \rangle_{eff}$ and the numerical values were obtained using the equation:

$$\left\langle \beta^2(2\omega) \right\rangle_{eff} = \frac{S_{GQD}(2\omega)}{S_{pNA}(2\omega)} \left[\left\{ \frac{N_{pNA}\left\langle \beta_{pNA}^2 \right\rangle + N_{mtOH}\left\langle \beta_{mtOH}^2 \right\rangle}{N_{GQD}} \right\} - \frac{N_{water}\left\langle \beta_{water}^2 \right\rangle}{N_{GQD}} \right], \quad (1)$$

where $mtOH$ refers to methanol, $\beta_{water} = 0.087 \times 10^{-30}$ esu [42], $N_{water} = 5.5 \times 10^{22}$ molecules/cm^3, $\beta_{p-NA} = 34 \times 10^{-30}$ esu [22,23], $N_{p-NA} = 1.0 \times 10^{19}$ molecules/cm^3,

$\beta_{mtOH} = 0.69 \times 10^{-30}$ esu [23], $N_{mtOH} = 1.50 \times 10^{22}$ molecules/cm^3. The concentration of GQDs was $N_{GQD} = 3.4 \times 10^{16}$ particles/cm^3 for both samples that were suspended in water. Considering the results obtained for $S_{GQD}(2\omega)/S_{pNA}(2\omega)$ (see Figure 4a), the effective values of $\langle \beta(2\omega) \rangle_{eff}$ were $(1.0 \pm 0.1) \times 10^{-27}$ esu and $(0.9 \pm 0.1) \times 10^{-27}$ esu for Acqua-Cyan and Acqua-Green, respectively. The quantity $\langle \beta^2(2\omega) \rangle_{eff}$ is an effective value contributed by GQDs with different number of nanosheets. Consequently, we considered the following approach to estimate the hyperpolarizability per nanosheet, $\langle \beta^2(2\omega) \rangle_{NS}$. First we considered the distribution of GQDs sizes shown in the Appendix A; then, assuming the nanosheet thickness of 1.5 nm, and a Gaussian distribution of GQDs sizes represented by $P(n) \propto \exp(-n^2/n_{av}^2)$, where n is the number of GQD nanoparticles, we obtain $10 \leq n \leq 60$ for Acqua-Cyan and $15 \leq n \leq 57$ for Acqua Green. The value of n_{av} is obtained from the AFM measurements given in the Appendix A. The value of $\langle \beta^2(2\omega) \rangle_{NS}$ is determined by $\langle \beta^2(2\omega) \rangle_{eff} = \langle \beta^2(2\omega) \rangle_{NS} \{\sum_1^n n\, P(n)/\sum_1^n P(n)\}$. Using these results, we could estimate the effective first hyperpolarizability per nanoshell, $\langle \beta(2\omega) \rangle_{eff}^{NS} \sim 10^{-28}$ esu, considering that each GQDs have five nanosheets with 0.3 nm of thickness.

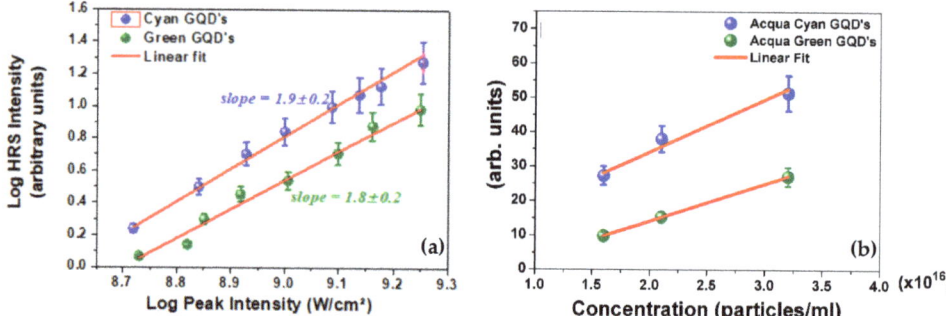

Figure 4. For better visibility, the data in blue (Acqua Cyan) have been shifted vertically. (a) HRS intensity signal, $S(2\omega) \propto I(2\omega)$, versus the laser intensity for both samples. The slopes of the straight lines were 1.9 ± 0.1 (Acqua-Cyan) and 1.8 ± 0.2 (Acqua-Green). (b) Dependence of $S(2\omega)$ with the concentration of GQDs per ml. The error bars represent 90% of confidence. The maximum peak intensity of the excitation source was 12 GW/cm^2.

The two-level model [41,43] was used to relate $\langle \beta(2\omega) \rangle_{eff}^{NS}$ with the first static hyperpolarizability per nanosheet, $\beta(0)^{NS}$, in each GQD. First, it is necessary to estimate the static hyperpolarizability, $\beta(0)$, for the samples. Then, considering the number of graphene sheets per GQD, $\beta(0)^{NS}$ is estimated. The following expression was obtained from [41,43].

$$\beta(0) = \left(1 - 4\frac{\omega^2}{\omega_{eg}^2}\right)\left(1 - \frac{\omega^2}{\omega_{eg}^2}\right)\langle \beta^2(2\omega) \rangle_{NS'} \qquad (2)$$

where $\omega_{eg} = 22{,}222$ cm^{-1} (Acqua-Cyan) and $\omega_{eg} = 19{,}608$ cm^{-1} (Acqua-Green). The frequency ω_{eg} correspond to the smallest band frequency in the absorption spectrum.

The results for $\beta(0)^{NS}$ are $(2.4 \pm 0.1) \times 10^{-28}$ esu and $(0.5 \pm 0.1) \times 10^{-28}$ esu for Acqua-Cyan and Acqua-Green, respectively. Notice that the value of $\langle \beta(2\omega) \rangle_{eff}^{NS}$ is approximately equal for both samples but the values of $\beta(0)^{NS}$ are different because of the different laser frequency detuning with respect to the first excited resonance frequency ω_{eg}.

We recall that imaging measurements have been performed using GQDs exploiting their large PL, but SHG microscopy presents an important advantage with respect to PL experiments because lasers operating in a large infrared wavelength range can be used since no resonance with excited states is required. This is a benefit because one may select the laser wavelength according to the transparency window of the biological system of interest. Obviously, care must be taken not to damage living cells and the underlying nanoparticles with too high intensities, as can often happen with thin films irradiated with pulsed lasers in the picosecond regime [44]. To characterize the origin of the HRS signal, polarization-resolved experiments were performed following the usual procedure [28–30]. HRS is a technique for investigating the arrangement of molecules in nanostructures at the nanoscale. The SHG output of randomly placed and oriented signals is polarization-dependent due to the properties of nanostructures that are determined by the arrangement of molecules at the microscopic level within these elements. The HRS study reveals important charge transfer differences within supramolecular systems in suspension, depending on the extent of electronic polarization induced by the constituting molecules [45]. The laser beam was linearly polarized and the angle γ, between the optical field and the vertical direction, varied from 0 to 2π. The polarization of the HRS signal, either in the vertical direction or parallel to the beam propagation axis, was measured. At this point, due to the insertion of a half-wave plate and a polarizer (see Figure 1), the laser intensity was ≈ 12 GW/cm^2. The results are shown in Figure 5 and, to fit the data, the following expression was used:

$$I_{HRS}^{\Gamma} = a^{\Gamma} \cos^4\gamma + b^{\Gamma} \cos^2\gamma \sin^2\gamma + c^{\Gamma} \sin^4\gamma + d^{\Gamma} \cos^3\gamma \sin\gamma + e^{\Gamma} \cos\gamma \sin^3\gamma \quad (3)$$

where I_{HRS}^{Γ} is the hyper—Rayleigh scattering intensity and Γ refers to the horizontal (H) or vertical (V) polarization. The coefficients a^{Γ}, b^{Γ}, c^{Γ}, d^{Γ}, e^{Γ}, were obtained from the fit. The measured values of d^{Γ} and e^{Γ} were too small compared to the other coefficients; therefore, these coefficients do not appear in Table 1. The multipolarity coefficients, $\varsigma^{\Gamma} = [1 - (a^{\Gamma} + c^{\Gamma})/b^{\Gamma}]$, were also determined and their values are indicated in Table 1.

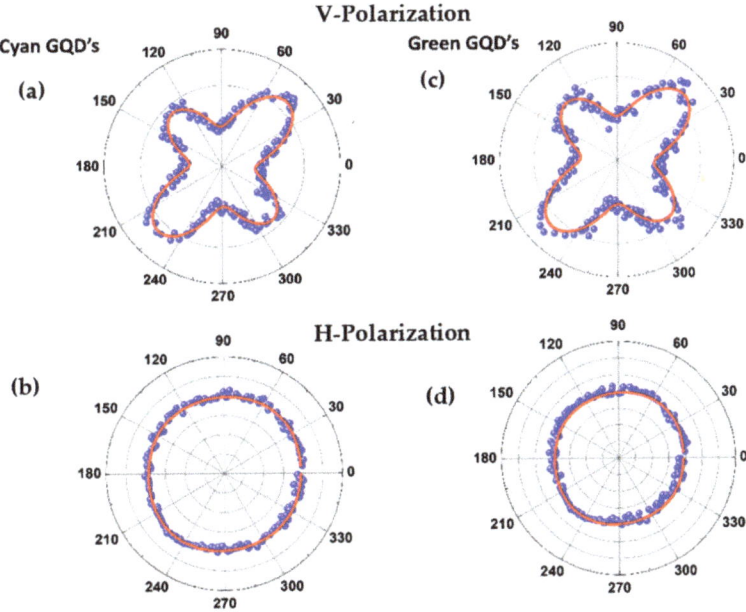

Figure 5. (a) Vertical polarization from Acqua-Cyan GQDs. (b) Horizontal polarization from Acqua-Cyan GQDs. (c) Vertical polarization from Acqua-Green GQDs. (d) Horizontal polarization from Acqua-Green GQDs. V and H polarizations are defined at the end of Section 2.

As mentioned before, the origin of the SHG rising from crystalline NPs presenting inversion symmetry is due to two factors: noncentrosymmetrical or centrosymmetrical shape. For nanoparticles presenting inversion symmetry and noncentrosymmetric shape, electric dipole surface second-harmonic response is allowed because of the surface of the particle, and the inversion symmetry of the bulk material is broken. In this case, retardation effects are not relevant for the SHG. On the other hand, for nanoparticles presenting inversion symmetry and centrosymmetrical shape, the retardation effects are important, and pure electric dipole contribution vanishes. The coefficient of multipolarity, ς^V, is applied to distinguish both contributions [39]. The case for pure centrosymmetric shape requires $a^V = c^V = 0$ and the case for the pure noncentrosymmetric shapes requires $b^V = a^V + c^V$. However, in real cases, the NPs are not perfect in shape or size. In this way, both the contribution of surface and volume may coexist. As one can see in Table 1, the values of ς^V for both samples suggest the dominant mechanism for harmonic generation is the field retardation in the NPs. This fact should be expected in our samples since there is no guarantee that the nanosheets are perfectly compacted and discordances should also occur between the sheets. As a result of such discordances, discontinuities in the electric field should occur in the NP volume and the quadrupole contributions become dominant. The negative value of the parameter ς^V can be interpreted as an effect from the nonspherical distribution of the GQD agglomerates [45].

Table 1. Multipolarity parameters for both GQDs samples.

Sample Parameters	Cyan GQD V Polarization	Green GQD V Polarization
a^Γ	0.87	0.79
b^Γ	0.56	0.59
c^Γ	0.14	0.19
ς^Γ	−0.81	−0.66

Ongoing research is focused on assessing the toxicity of graphene-family nanoparticles, with specific attention to GQDs [46]. Factors such as particle size, synthesis methods, and chemical doping play a role in determining their toxicity, both in vivo and in terms of cytotoxicity [47]. Some researchers argue that GQDs, composed primarily of organic materials, exhibit low toxicity and high biocompatibility, providing an advantage over semiconductor quantum dots. In vitro studies using cell cultures have shown minimal bad effects of GQDs on human cell viability [48]. Fluorescence imaging, a non-radioactive method, allows the visualization of morphological details in various biological specimens, from living cells to animals. Notably, GQDs, unlike many semiconductor quantum dots containing heavy metals, consist of graphene lattices containing light elements. Because they are primarily made of carbon, the most abundant element in biological systems, GQDs are generally considered biocompatible.

The analysis of beta per nanosheet is based on the number of nanosheets only, without considering their relative orientation and the fine structure of the scatterers. While our study sheds light on the polarization-dependent behavior of GQDs, further investigation is needed to elucidate the underlying mechanisms governing this phenomenon and to explore potential applications in detail. Future studies could delve into the specific factors influencing the polarization dependency, such as GQD size, shape, and surface functionalization. Additionally, exploring the potential integration of GQDs into practical devices and investigating their performance in real-world conditions would be instrumental.

4. Summary and Conclusions

In summary, we reported the incoherent second harmonic generation by aqueous suspensions of graphene quantum dots excited at 1064 nm. The hyper-Rayleigh scattering technique was applied for the determination of the first hyperpolarizability associated with the GQDs and the results indicate $\langle \beta(2\omega) \rangle_{eff}^{NS} \sim 10^{-28}$ esu per nanosheet. The spectra of the scattered light (centered at the second harmonic frequency) as well as their fast temporal behavior prove that the signals detected are due to the second harmonic generation and not due to luminescence induced by two-photon absorption. From the hyper-Rayleigh scattering data, using a two-level model, we could determine the static hyperpolarizability per nanosheet, $\beta(0)^{NS}$. The values obtained were $(2.4 \pm 0.1) \times 10^{-28}$ esu for Acqua-Cyan and $(0.5 \pm 0.1) \times 10^{-28}$ esu for Acqua-Green. In the future, could we develop a straightforward theoretical model for predicting the second-order coefficients of optical nonlinearity, much like we can for third-order optical nonlinear coefficients? Theoretical forecasts for the nonlinear refractive index in infrared glasses are achievable [49]. The order of magnitude of the GQDs hyperpolarizabilities indicates that the two samples studied have potential for application in the microscopy of biological systems. Of course, the possible nontoxicity of the GQDs to living cells is also a relevant characteristic in comparison with many semiconductor quantum dots being used.

Author Contributions: Conceptualization, C.B.d.A.; methodology, C.B.d.A.; software, M.L.S.-N., G.B. and R.B.-S.; validation, C.B.d.A., M.L.S.-N. and R.B.-S.; formal analysis, C.B.d.A.; investigation, M.L.S.-N., R.B.-S.; resources, C.B.d.A., M.L.S.-N., G.B. and R.B.-S.; data curation, M.L.S.-N. and R.B.-S.; writing—original draft preparation, C.B.d.A., M.L.S.-N. and R.B.-S.; writing—review and editing, M.L.S.-N., G.B. and R.B.-S.; supervision, C.B.d.A.; project administration, C.B.d.A.; funding acquisition, C.B.d.A., G.B. All authors have read and agreed to the published version of the manuscript.

Funding: This research was funded by the Brazilian agencies Conselho Nacional de Desenvolvimento Científico e Tecnológico (CNPq), Fundação de Amparo à Ciência e Tecnologia do Estado de Pernambuco (FACEPE), Coordenação de Aperfeiçoamento de Pessoal de Ensino Superior (CAPES), and the University of Angers: AAP MIR 2022. This work was performed in the framework of the National Institute of Photonics (INFO) project.

Informed Consent Statement: Not applicable.

Data Availability Statement: Data underlying the results presented in this paper are not publicly available at this time but may be obtained from the authors upon reasonable request.

Acknowledgments: G.B. would like to acknowledge the financial support from the NNN-TELECOM Program, region des Pays de la Loire, contract n°: 2015 09036.

Conflicts of Interest: The authors declare no conflict of interest.

Appendix A

Characterization of the GQDs

The GQD dimensions were measured by Atomic Force Microscope (AFM) measurements. Figure A1 shows the AFM image for Acqua-Cyan and Acqua-Green GQDs.

From the AFM images, it was possible to make a histogram of the diameter and height of the GQDs, as shown below.

Figure A2 shows the histogram for the diameter and height of Acqua-Cyan and Acqua-Green GQDs. The log-normal fit indicates a maximum distribution for a diameter of 4 nm and height of 52 nm for Acqua-Cyan GQDs. For the Acqua-Green GQDs, the size distribution peaked for a diameter of 3.9 nm and a height of 46 nm.

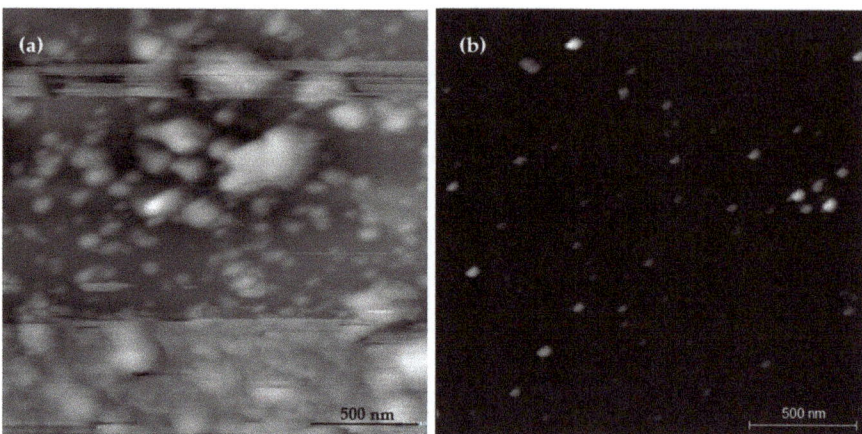

Figure A1. AFM images: (**a**) Acqua-Cyan GQDs; (**b**) Acqua-Green GQDs. The scale is 500 nm for both images.

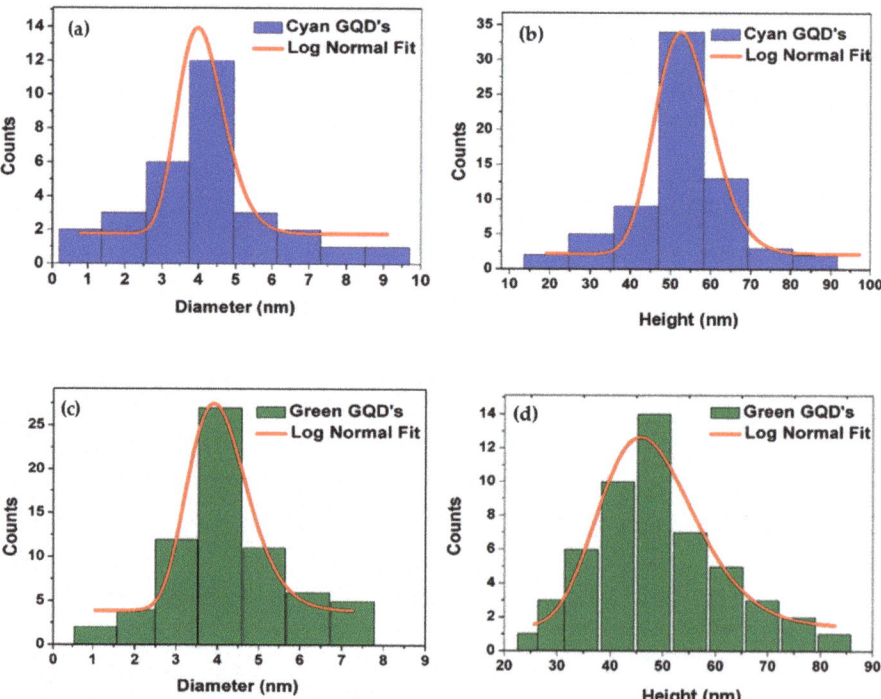

Figure A2. (**a**) Diameter and (**b**) Height of Acqua-Cyan GQDs; (**c**) Diameter and (**d**) Height of Acqua-Green GQDs.

Using the size distribution, we could calculate the concentration of GQDs per cm^3. The value obtained was 0.9×10^{15} GQDs/cm^3 for Acqua-Cyan and 1×10^{15} GQDs/cm^3 for Acqua-Green.

Two-photon absorption-induced luminescence (2PL) with excitation at 850 nm was performed for both GQDs. The Figure A3a,b show the 1PL (solid lines) and 2PL (dots) spectra for both samples.

In Figure A3c,d we show the temporal behavior of 2PL, with a fluorescence lifetime of 19 ns for Acqua-Cyan GQDs and 26 ns for Green GQDs.

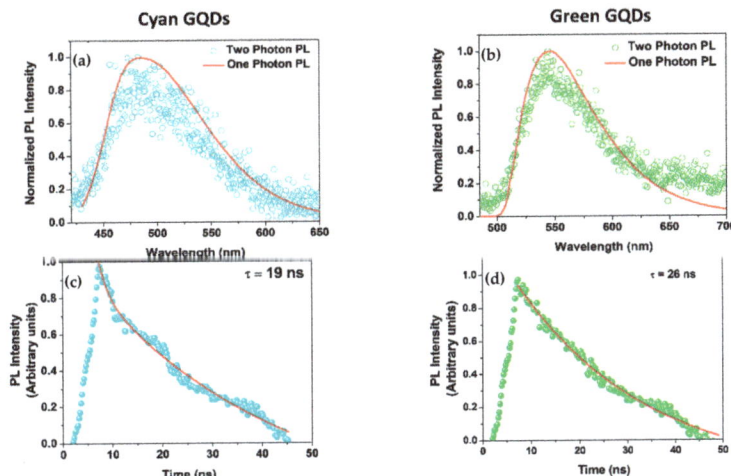

Figure A3. Normalized 1PL and 2PL Intensity (**a**) Acqua-Cyan GQDs and (**b**) Acqua Green GQDs. 2PL lifetime for (**c**) Acqua-Cyan GQDs and (**d**) Acqua-Green GQDs.

References

1. Tang, L.; Ji, R.; Cao, X.; Lin, J.; Jiang, H.; Li, X.; Teng, K.S.; Luk, C.M.; Zeng, S.; Hao, J.; et al. Deep Ultraviolet Photoluminescence of Water-Soluble Self-Passivated Graphene Quantum Dots. *ACS Nano* **2012**, *6*, 5102–5110. [CrossRef]
2. Luo, Z.; Lu, Y.; Somers, L.A.; Johnson, A.C. High Yield Preparation of Macroscopic Graphene Oxide Membranes. *J. Am. Chem. Soc.* **2009**, *131*, 898–899. [CrossRef]
3. Bacon, M.; Bradley, S.J.; Nann, T. Graphene Quantum Dots. *Part. Part. Syst. Charact.* **2014**, *31*, 415–428. [CrossRef]
4. Chen, W.; Lu, G.; Hu, W.; Li, D.; Chen, S.; Dai, Z. Synthesis and Applications of Graphene Quantum Dots: A Review. *Nanotechnol. Rev.* **2018**, *7*, 157–185. [CrossRef]
5. Wettstein, C.M.; Bonafé, F.P.; Oviedo, M.B.; Sánchez, C.G. Optical Properties of Graphene Nanoflakes: Shape Matters. *J. Chem. Phys.* **2016**, *144*, 224305. [CrossRef] [PubMed]
6. Feng, X.; Qin, Y.; Liu, Y. Size and Edge Dependence of Two-Photon Absorption in Rectangular Graphene Quantum Dots. *Opt. Express* **2018**, *26*, 7132–7139. [CrossRef]
7. Cayuela, A.; Soriano, M.L.; Carrillo-Carrion, C.; Valcárcel, M. Semiconductor and Carbon-Based Fluorescent Nanodots: The Need for Consistency. *Chem. Commun.* **2006**, *52*, 1311–1326. [CrossRef]
8. Feng, X.; Li, Z.; Li, X.; Liu, Y. Giant Two-Photon Absorption in Circular Graphene Quantum Dots in Infrared Region. *Sci. Rep.* **2016**, *6*, 33260. [CrossRef]
9. Zhao, C.; Song, X.; Liu, Y.; Fu, Y.; Ye, L.; Wang, N.; Wang, F.; Li, L.; Mohammadniaei, M.; Zhang, M.; et al. Synthesis of graphene quantum dots and their applications in drug delivery. *J. Nanobiotechnol.* **2020**, *18*, 142. [CrossRef] [PubMed]
10. Kalkal, A.; Kadian, S.; Pradhan, R.; Manik, G.; Packirisamy, G. Recent advances in graphene quantum dot-based optical and electrochemical (bio) analytical sensors. *Mater. Adv.* **2021**, *2*, 5513–5541. [CrossRef]
11. Iannazzo, D.; Ziccarelli, I.; Pistone, A. Graphene Quantum Dots: Multifunctional Nanoplatforms for Anticancer Therapy. *J. Mater. Chem. B* **2017**, *5*, 6471–6489. [CrossRef] [PubMed]
12. Thakur, M.; Kumawat, M.K.; Srivastava, R. Multifunctional Graphene Quantum Dots for Combined Photothermal and Photodynamic Therapy Coupled with Cancer Cell Tracking Applications. *RSC Adv.* **2017**, *7*, 5251–5261. [CrossRef]
13. Bak, S.; Kim, D.; Lee, H. Graphene Quantum Dots and Their Possible Energy Applications: A Review. *Curr. Appl. Phys.* **2016**, *16*, 1192–1201. [CrossRef]
14. Li, X.; Rui, M.; Song, J.; Shen, Z.; Zeng, H. Carbon and Graphene Quantum Dots for Optoelectronic and Energy Devices: A Review. *Adv. Funct. Mater.* **2015**, *25*, 4929–4947. [CrossRef]
15. Qi, X.; Liu, H.; Guo, W.; Lin, W.; Lin, B.; Jin, Y.; Deng, X. New opportunities: Second harmonic generation of boron-doped graphene quantum dots for stem cells imaging and ultraprecise tracking in wound healing. *Adv. Funct. Mater.* **2019**, *29*, 1902235. [CrossRef]

16. Li, H.P.; Bi, Z.T.; Xu, R.F.; Han, K.; Li, M.X.; Shen, X.P.; Wu, Y.X. Theoretical Study on Electronic Polarizability and Second Hyperpolarizability of Hexagonal Graphene Quantum Dots: Effects of Size, Substituent, and Frequency. *Carbon* **2017**, *122*, 756–760. [CrossRef]
17. Cheng, J.L.; Vermeulen, N.; Sipe, J.E. Numerical Study of the Optical Nonlinearity of Doped and Gapped Graphene: From Weak to Strong Field Excitation. *Phys. Rev. B* **2015**, *92*, 235307. [CrossRef]
18. Liu, Q.; Guo, B.; Rao, Z.; Zhang, B.; Gong, J.R. Strong Two-Photon-Induced Fluorescence from Photostable, Biocompatible Nitrogen-Doped Graphene Quantum Dots for Cellular and Deep-Tissue Imaging. *Nano Lett.* **2013**, *13*, 2436–2441. [CrossRef]
19. Boudebs, G.; Cherukulappurath, S. Nonlinear refraction measurements in presence of nonlinear absorption using phase object in a 4f system. *Opt. Commun.* **2005**, *250*, 416–420. [CrossRef]
20. Yamijala, S.S.; Mukhopadhyay, M.; Pati, S.K. Linear and Nonlinear Optical Properties of Graphene Quantum Dots: A Computational Study. *J. Phys. Chem. C* **2015**, *119*, 12079–12087. [CrossRef]
21. Zhou, Z.J.; Li, X.P.; Ma, F.; Liu, Z.B.; Li, Z.R.; Huang, X.R.; Sun, C.C. Exceptionally Large Second-Order Nonlinear Optical Response in Donor–Graphene Nanoribbon–Acceptor Systems. *Chem. Eur. J.* **2011**, *17*, 2414–2419. [CrossRef]
22. Zhou, Z.J.; Liu, Z.B.; Li, Z.R.; Huang, X.R.; Su, C.C. Shape Effect of Graphene Quantum Dots on Enhancing Second-Order Nonlinear Optical Response and Spin Multiplicity in NH_2-GQD-NO_2 System. *J. Phys. Chem. C* **2011**, *115*, 16282–16286. [CrossRef]
23. Clays, K.; Persoons, A. Hyper-Rayleigh Scattering in Solution. *Phys. Rev. Lett.* **1991**, *66*, 2980–2983. [CrossRef] [PubMed]
24. Clays, K.; Olbrechts, G.; Munters, T.; Persoons, A.; Kim, O.-K.; Choi, L.-S. Enhancement of the Molecular Hyperpolarizability by a Supramolecular Amylose–Dye Inclusion Complex, Studied by Hyper-Rayleigh Scattering with Fluorescence Suppression. *Chem. Phys. Lett.* **1998**, *293*, 337–342. [CrossRef]
25. Rodriguez, M.B.; Shelton, D.P. What is Measured by Hyper Rayleigh Scattering from a Liquid? *J. Chem. Phys.* **2018**, *148*, 134504. [CrossRef]
26. Collins, J.T.; Rusimova, K.R.; Hooper, D.C.; Jeong, H.H.; Ohnoutek, L.; Pradaux-Caggiano, F.; Valev, V.K. First Observation of Optical Activity in Hyper-Rayleigh Scattering. *Phys. Rev. X* **2019**, *9*, 011024. [CrossRef]
27. Barbosa-Silva, R.; Nogueira, M.A.M.; Souza, H.D.S.; Lira, B.F.; de Athaide-Filho, P.F.; de Araújo, C.B. First Hyperpolarizability of 1,3-Thiazolium-5-Thiolates Mesoionic Compounds. *J. Phys. Chem. C* **2019**, *123*, 677–683. [CrossRef]
28. El Harfouch, Y.; Benichou, E.; Bertorelle, F.; Russier-Antoine, I.; Jonin, C.; Lascoux, N.; Brevet, P.F. Hyper-Rayleigh Scattering from Gold Nanorods. *J. Phys. Chem. C* **2013**, *118*, 609–616. [CrossRef]
29. Russier-Antoine, I.; Lee, H.J.; Wark, A.W.; Butet, J.; Benichou, E.; Jonin, C.; Brevet, P.F. Second Harmonic Scattering from Silver Nanocubes. *J. Phys. Chem. C* **2018**, *122*, 17447–17455. [CrossRef]
30. Khebbache, N.; Maurice, A.; Djabi, S.; Russier-Antoine, I.; Jonin, C.; Skipetrov, S.E.; Brevet, P.F. Second-Harmonic Scattering from Metallic Nanoparticles in a Random Medium. *ACS Photon.* **2017**, *4*, 262–267. [CrossRef]
31. Joulaud, C.; Mugnier, Y.; Djanta, G.; Dubled, M.; Marty, J.C.; Galez, C.; Wolf, J.P.; Bonacina, L.; Dantec, R.L. Characterization of the Nonlinear Optical Properties of Nanocrystals by Hyper Rayleigh Scattering. *J. Nanobiotechnol.* **2013**, *11* (Suppl. S1), S8. [CrossRef]
32. Forcherio, G.T.; Riporto, J.; Dunklin, J.R.; Mugnier, Y.; Dantec, R.L.; Bonacina, L.; Roper, D.K. Nonlinear Optical Susceptibility of Two-Dimensional WS_2 Measured by Hyper Rayleigh Scattering. *Opt. Lett.* **2017**, *42*, 5018–5021. [CrossRef] [PubMed]
33. Valdez, E.; de Araújo, C.B.; Lipovskii, A.A. Second Harmonic Scattered Light from a Transparent Glass-Ceramic Containing Sodium Niobate Nanocrystals. *Appl. Phys. Lett.* **2006**, *89*, 031901. [CrossRef]
34. Rodriguez, E.V.; de Araújo, C.B.; Brito-Silva, A.M.; Ivanenko, V.I.; Lipovskii, A.A. Hyper-Rayleigh Scattering from $BaTiO_3$ and $PbTiO_3$ Nanocrystals. *Chem. Phys. Lett.* **2009**, *467*, 335–338. [CrossRef]
35. Barbosa-Silva, R.; Silva, J.F.; Rocha, U.; Jacinto, C.; de Araújo, C.B. Second-Order Nonlinearity of $NaNbO_3$ Nanocystals with Orthorhombic Crystalline Structure. *J. Lumin.* **2019**, *211*, 121–126. [CrossRef]
36. Dadap, J.I.; Shan, J.; Heinz, T.F. Theory of optical second-harmonic generation from a sphere of centrosymmetric material: Small-particle limit. *J. Opt. Soc. Am. B* **2004**, *7*, 1328–1347. [CrossRef]
37. Bachelier, G.; Russier-Antoine, I.; Benichou, E.; Jonin, C.; Brevet, P.F. Multipolar second-harmonic generation in noble metal nanoparticles. *J. Opt. Soc. Am. B* **2008**, *25*, 955–960. [CrossRef]
38. Nappa, J.; Revillod, G.; Russier-Antoine, I.; Benichou, E.; Jonin, C.; Brevet, P.F. Electric dipole origin of the second harmonic generation of small metallic particles. *Phys. Rev. B* **2005**, *71*, 165407. [CrossRef]
39. Brevet, P.F. Second harmonic generation in nanostructures. In *Handbook of Nanoscale Optics and Electronics*; Wiederrecht, G.P., Ed.; Elsevier: Amsterdam, The Netherlands, 2010; pp. 351–381.
40. Available online: https://www.strem.com/uploads/resources/documents/graphene_quantum_dots_-_dotz_nano.pdf (accessed on 15 November 2023).
41. Marder, S.R.; Beratan, D.N.; Cheng, L.T. Approaches for Optimizing the First Electronic Hyperpolarizability of Conjugated Organic Molecules. *Science* **1991**, *252*, 103–106. [CrossRef]
42. Pauley, M.A.; Guan, H.W.; Wang, C.H.; Jen, A.K.Y. Determination of First Hyperpolarizability of Nonlinear Optical Chromophores by Second Harmonic Scattering Using an External Reference. *J. Chem. Phys.* **1996**, *104*, 7821–7829. [CrossRef]
43. Chemla, D.S. *Nonlinear Optical Properties of Organic Molecules and Crystals*; Academic Press: New York, NY, USA, 1987.
44. Fedus, K.; Boudebs, G.; de Araujo, C.B.; Cathelinaud, M.; Charpentier, F.; Nazabal, V. Photoinduced effects in thin films of $Te_{20}As_{30}Se_{50}$ glass with nonlinear characterization. *Appl. Phys. Lett.* **2009**, *94*, 6. [CrossRef]

45. Revillod, G.; Duboisset, J.; Russier-Antoine, I.; Benichou, E.; Jonin, C.; Brevet, P.F. Second Harmonic Scattering of Molecular Aggregates. *Symmetry* **2021**, *13*, 206. [CrossRef]
46. Ou, L.; Song, B.; Liang, H.; Liu, J.; Feng, X.; Deng, B.; Sun, T.; Shao, L. Toxicity of graphene-family nanoparticles: A general review of the origins and mechanisms. *Part. Fibre Toxicol.* **2016**, *13*, 57. [CrossRef] [PubMed]
47. Wang, S.; Cole, I.S.; Li, Q. The toxicity of graphene quantum dots. *RSC Adv.* **2016**, *6*, 89867–89878. [CrossRef]
48. Ghosh, S.; Sachdeva, B.; Sachdeva, P.; Chaudhary, V.; Rani, G.M.; Sinha, J.K. Graphene quantum dots as a potential diagnostic and therapeutic tool for the management of Alzheimer's disease. *Carbon Lett.* **2022**, *32*, 1381–1394. [CrossRef]
49. Fedus, K.; Boudebs, G.; Coulombier, Q.; Troles, J.; Zhang, X.H. Nonlinear characterization of GeS_2–Sb_2S_3–CsI glass system. *J. Appl. Phys.* **2010**, *107*, 023108. [CrossRef]

Disclaimer/Publisher's Note: The statements, opinions and data contained in all publications are solely those of the individual author(s) and contributor(s) and not of MDPI and/or the editor(s). MDPI and/or the editor(s) disclaim responsibility for any injury to people or property resulting from any ideas, methods, instructions or products referred to in the content.

Communication

Cubic Nonlinearity of Graphene-Oxide Monolayer

Tikaram Neupane [1], Uma Poudyal [1], Bagher Tabibi [2], Wan-Joong Kim [3] and Felix Jaetae Seo [2,*]

[1] Department of Chemistry and Physics, The University of North Carolina at Pembroke, Pembroke, NC 28372, USA; tikaram.neupane@uncp.edu (T.N.); uma.poudyal@uncp.edu (U.P.)
[2] Advanced Center for Laser Science and Spectroscopy, Department of Physics, Hampton University, Hampton, VA 23668, USA; bagher.tabibi@gmail.com
[3] K1 Solution R&D Center, Geumcheon-gu, Seoul 08591, Republic of Korea; kokwj@daum.net
* Correspondence: jaetae.seo@hamptonu.edu

Abstract: The cubic nonlinearity of a graphene-oxide monolayer was characterized through open and closed z−scan experiments, using a nano-second laser operating at a 10 Hz repetition rate and featuring a Gaussian spatial beam profile. The open z−scan revealed a reverse saturable absorption, indicating a positive nonlinear absorption coefficient, while the closed z−scan displayed valley-peak traces, indicative of positive nonlinear refraction. This observation suggests that, under the given excitation wavelength, a two-photon or two-step excitation process occurs due to the increased absorption in both the lower visible and upper UV wavelength regions. This finding implies that graphene oxide exhibits a higher excited-state absorption cross-section compared to its ground state. The resulting nonlinear absorption and nonlinear refraction coefficients were estimated to be approximately ~2.62×10^{-8} m/W and 3.9×10^{-15} m^2/W, respectively. Additionally, this study sheds light on the interplay between nonlinear absorption and nonlinear refraction traces, providing valuable insights into the material's optical properties.

Keywords: one-photon transition; two-photon transition; 2D materials; nonlinear absorption; nonlinear refraction

1. Introduction

The successful extraction of a graphene single layer through mechanical cleavage in 2004 has ignited a wave of research interest in the fascinating world of two-dimensional (2D) materials [1]. Graphene's exceptional Kerr effect and high nonlinear absorption coefficient make it a highly promising material for various applications in the field of optoelectronics [2,3]. However, the high nonlinear absorption coefficient was accompanied by a zero bandgap through the strong two-photon absorption (TPA) process, which might not only be from the two-step excitation but also due to an undesired free-carrier absorption (FCA) and free-carrier dispersion (FCD) [4]. On the other hand, graphene oxide (GO) has been recognized as one of the growing two-dimensional materials due to its broadband optical effect through the tunable bandgap [5,6]. The content and location of oxygen-containing groups obviously influence the optical and electronic properties [7,8]. These reduction processes transform GO from an insulator to a semiconductor and to a metal-like state, in the form of graphene. In addition, GO is a hydrophilic and water-soluble material due to the existence of the oxygen-containing group which makes the fabrication easier.

The third-order (cubic) nonlinear optical properties of GO are stable under high-power illumination [9], which makes GO a strong candidate for the optoelectronic applications such as pulse compression, mode-locking to Q-switching, optical limiting (OL), and all-optical switching [5,10–23]. The study of third-order optical nonlinearity, which includes the coefficient of nonlinear absorption (NLA), nonlinear refraction (NLR), and their polarity as well, reveals its prospective technical applications in optoelectronics. For instance, materials exhibiting negative nonlinear absorption (NLA) coefficients, characterized by saturable absorption (SA), find application as crucial Q-switching elements in lasers [24]. Conversely, materials featuring positive NLA coefficients, demonstrating reverse saturable absorption (RSA), are well suited for applications such as two-photon microscopy and optical limiters [25]. To study such an enthralling phenomenon, the "z−scan technique" has been used to investigate both the polarity and magnitude of NLA and NLR coefficient [16,26]. Kang et al., revealed that GO exhibits strong and broadband nonlinear optical (NLO) properties for the optical power limiting in a wide spectral range of wavelengths [27]. It obviously meets the demands for emerging photonic applications to overcome the OL behavior at certain wavelengths via metallic nanomaterials (zinc ferrite nanoparticles [28], gold clusters [29]). In addition, the NLO response in a wide spectral range exhibits SA at the short wavelength and RSA at the longer wavelength, which may result in a varying magnitude of third-order susceptibility based on the wavelength-dependent nonlinear optical transition [27]. The magnitude and polarity of third-order nonlinearity are modified using excitation processes such as resonant and non-resonant nonlinear processes. As an example, resonant excitation yields a relatively greater magnitude of nonlinearity compared to non-resonant excitation. However, it is worth noting that resonant excitation demands a longer duration within the nonlinear process compared to its non-resonant counterpart [30]. The literature review divulges that the few layers of GO dispersion and reduced GO (rGO) are good candidates for the optical power limiting at the 532 nm wavelength [31–34]. Also, GO nanosheets dispersed in DI water demonstrated the broadband NLO and its prospective application in optical power limiting at the 1064 nm wavelength [35]. The tunable OL properties of GO in ethanol solution were studied at 1550 nm as well [36]. In moving towards the application of GO as an optical power limiter, the characterization of third-order optical susceptibility (χ^3) is a pivotal parameter that is directly related to the coefficients of NLA and NLR. Additionally, the imaginary χ^3 articulates the information about the SA and RSA to identify the possible application as an optical power limiter.

Ebrahimi et al. reported the imaginary (Im) χ^3 value of GO in ethanol at 532 nm, which was 2.17×10^{-14} m^2/V^2 [37]. The values of (Im) χ^3 change with GO dispersion in different solvents at 532 nm that varied from 1.0×10^{-19} to 5.1×10^{-19} m^2/V^2 [31]. In addition, Khanzadeh revealed that the χ^3 of GO was 5.12×10^{-16} m^2/V^2. At the same wavelength, Kang et al. demonstrated that the χ^3 value of GO is 8.97×10^{-18} m^2/V^2 [27]. It implies that the different magnitude of χ^3 arises from the oxygen content in the sample, the types of solvent used, and the applied wavelength [38]. Our estimation of χ^3 at 532 nm is one order higher than the one investigated using a continuous wavelength from 450 to 750 nm as per the literature review [27]. Furthermore, this study exhibits a positive nonlinear absorption coefficient, indicating that the excited state absorption cross-section surpasses that of the ground state. This characteristic makes it a promising candidate for utilization as an optical power limiter. It is worth noting that the highly intense laser beam employed in the nonlinear experiments may introduce a thermal effect into the results obtained from the z−scan [39]. Therefore, this article has provided a comprehensive analysis of both the polarity and magnitude of the nonlinear absorption and nonlinear refraction coefficients of graphene oxide (GO) in DI water. These measurements were conducted using a nanosecond laser emitting at a visible wavelength of 532 nm in a z−scan setup.

2. Materials and Methods

The GO nanoflakes in an aqueous solution were purchased from a graphene supermarket, Ronkonkoma, NY [40]. According to the vendor's specifications, the GO product consists of more than 80% monolayers with nanoflakes ranging in size from 0.5 to 5 microns (https://www.graphene-supermarket.com/, accessed on 10 February 2023). The experiment utilized diluted GO, which was prepared using a standard 10 mm cuvette with a capacity of approximately 3.5 mL for dilution. The process involved mixing approximately 1.75 mL of the GO sample, which had a concentration of 6 g/L, with an equal volume of DI water. The third-order nonlinear optical properties of GO nanoflakes were characterized using the z−scan method in a 1 mm quartz cuvette. The excitation source for the z−scan was a pulsed laser at 532 nm, 10-Hz repetition rate, and ~6–ns temporal pulse width with the Gaussian beam [41]. The effective focal length of the focusing lens for the z–scan was ~125 mm. The radius of the beam waist (w_0) at the focal point was ~15.1 μm. The Rayleigh length ($z_0 = kw_0^2/2$) was ~2.02 mm, which was larger than the sample thickness (1 mm). The Gaussian beam at the focusing lens was ~2.8 mm at FWHM and was prepared using the two-irises method [41]. To ensure the integrity of our experimental setup, we meticulously eliminated all potential optical nonlinearities originating from electronic devices and other optical components by conducting experiments within the system's linear range. Additionally, we performed z–scan experiments using only the base solvent to eliminate any undesired supplementary nonlinear effects from the sample. For validation purposes, we employed the nonlinear refraction of CS_2 to verify the experimental setup through the closed z–scan technique [41]. The schematic diagrams of open and closed z–scan setups are depicted in Figure 1a and 1b, respectively [26], and have been adapted with permission from the previous publication [41].

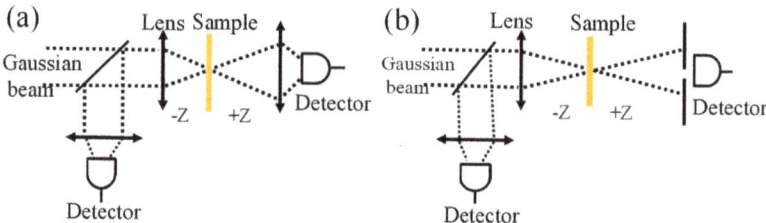

Figure 1. Schematic diagram of open (**a**) and closed (**b**) z–scan setups for characterizing the magnitude and polarity of nonlinear absorption and nonlinear refraction, respectively [41].

3. Results and Discussion

In the linear absorption spectrum of GO nanoflakes in DI water, the absorption peaks are evident at around 230 nm and 300 nm, with their spectral tails extending into the visible region, as illustrated in Figure 2. The p orbitals of carbon can be combined in two ways: in-phase, resulting in bonding combinations, and out of phase, leading to anti-bonding combinations. This results in the formation of π and π^* orbitals, with the π orbital having lower energy compared to the π^* orbital. Consequently, this energy difference facilitates photon-induced transitions between the π and π^* orbitals. The initial peak corresponds to the π–π^* transition of the C=C bond, while the subsequent peak corresponds to the n–π^* transition involving the carbonyl (C=O) bonds of GO [42].

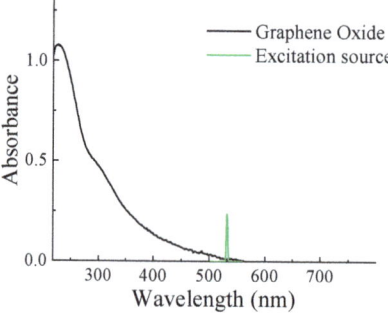

Figure 2. The absorption spectrum of GO nanoflakes in DI water. The green spectrum indicates the laser excitation used for nonlinear optical characterization.

The nonlinear absorption behavior of GO nanoflakes in DI water was characterized through an open z–scan technique, revealing normalized nonlinear transmittance as a function of sample position (z) across various peak excitation intensities, namely ~6.4 GW/cm^2,

3.5 GW/cm², 1.9 GW/cm², 1.1 GW/cm², and 0.1 GW/cm² at the focal plane (refer to Figure 3). These nonlinear transmittance traces distinctly exhibit a reverse saturable absorption (RSA) pattern, indicative of a positive nonlinear absorption coefficient [14]. This suggests that the absorption cross-section in the excited state surpasses that of the ground state. At lower wavelength regions, there is a notable increase in absorption, indicating the potential occurrence of two-photon excitation or two-photon absorption within the framework of optical nonlinearity. The normalized nonlinear transmittance via an open z−scan for a Gaussian beam is governed by [16].

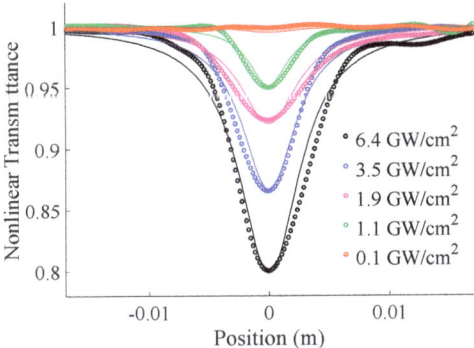

Figure 3. Normalized nonlinear transmittance as a function of the sample position (z) through the open z−scan technique, encompassing various applied peak intensities.

$$T(z, S = 1) = \sum_{m=0}^{\infty} \frac{\left(\frac{-q}{1+(z/z_o)^2}\right)^m}{(1+m)^{3/2}} \qquad (1)$$

where z is the sample position, $S = 1 - \exp(-2r_a^2/w_0^2) = 1$ is the unit linear transmittance indicating no aperture in front of the detector in the open z−scan, r_a is the radius of a finite aperture, $q(r,z,t) = \beta I_0 L_{eff} < 1$ is the requirement for the open z−scan, which results in the negligible nonlinear phase distortion of $\Delta\Psi_o(t) = \beta I_0(t) L_{eff}/2$, $L_{eff} = (1 - \exp(-\alpha_0 L))/\alpha_0 L$ is the effective sample length, L is the sample thickness, I_o is the applied peak intensity, α_o is the linear absorption coefficient, and β is the nonlinear absorption coefficient. By fitting it to the model Equation (1), we estimate the nonlinear absorption coefficient of GO to be approximately ~2.62 × 10⁻⁸ m/W.

Furthermore, the normalized nonlinear refraction traces are characterized by closed z−scan techniques as shown in Figure 4a. It satisfied the far-field condition of an aperture (d~2.0 >> z_o). The radius (r_a) of a finite aperture is selected at ~0.75 mm to satisfy the linear transmittance of the finite aperture (S)~0.01 < 1 condition [41]. The normalized transmittance, depicted as a function of the sample position, showcases the presence of valley-peak traces, as exemplified in Figure 4. It suggests the self-focusing characteristics or positive nonlinear refraction observed in GO nanoflakes. The theoretical model of normalized nonlinear transmittance using a closed z−scan with a Gaussian beam is [26,43].

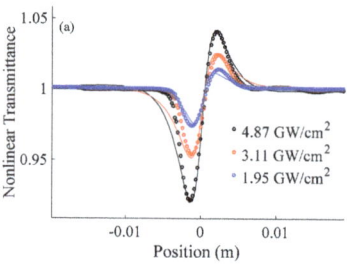
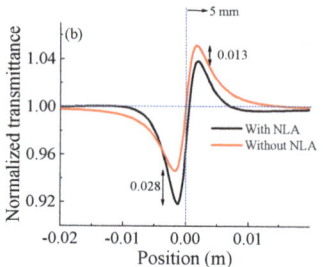

Figure 4. (**a**) Experimental normalized transmittance as a function of sample position (z) obtained through the closed z−scan technique, along with their corresponding fitting traces. (**b**) A comparative analysis of traces with and without the nonlinear absorption (NLA) coefficient at an applied intensity of 4.87 GW/cm^2.

$$T(S \ll 1) = 1 - \frac{4\Delta\varphi_0 x + q(3 + x^2)}{(1 + x^2)(9 + x^2)} - \frac{4\Delta\varphi_0^2(5 - 3x^2) - 8\Delta\varphi_0 qx(9 + x^2) - q^2(40 + 17x^2 + x^4)}{(1 + x^2)(9 + x^2)(25 + x^2)} + \ldots \quad (2)$$

where w_a is the radius of beam waist at the focal plane, $q(r, z, t) = \beta I_0 L_{eff} < 1$, $\Delta\varphi_o = k\gamma I_0 L_{eff} < 1$ is the phase distortion for the symmetric peak-valley nonlinear transmittance trace, γ is the nonlinear refraction coefficient, and $x = -(1/z_o)(z + (z_o^2 + z^2)/d - z) \sim z/z_o$ for the far-field condition of an aperture ($d \gg z_o$), where d is the distance between the focal plane and the aperture. The nonlinear refraction coefficient (γ) is found to be $\sim 3.9 \times 10^{-15}$ m^2/W from fitting with model Equation (2). Looking into the valley-peak transmittance, the valley is much deeper (~ 0.028 $T_{v\text{-}p}$) than the valley-peak symmetry. This implies that absorption plays a considerable role in nonlinear refraction [44].

The NLR curve governed by Equation (2) results in transmittance variation between the normalized peak and valley $T_{p\text{-}v} = 0.4 |\Delta\phi_o|$ and a peak–valley separation of $z_{p\text{-}v} = 1.79\, z_0$. Also, $\Delta T_{p\text{-}v}$ depends on the magnitude of the nonlinear phase shift ($\Delta\phi_o$) and the linear transmittance of the finite aperture (S). For a smaller phase shift ($\Delta\phi_o \leq \pi$), which resembles this experiment, the $\Delta T_{p\text{-}v}$ follows the equation within ±2% variation [16].

$$\Delta T_{p-v} \approx 0.406(1 - S)^{0.25} |\Delta\phi_o| \quad (3)$$

For an applied intensity, the phase shift is constant. Therefore, this article is focused on investigating the effect of S on $\Delta T_{p\text{-}v}$ using the estimated nonlinear refraction coefficient for three different applied peak intensities.

As depicted in Figure 5, it is evident that the magnitude of the peak-valley difference ($\Delta T_{p\text{-}v}$) decreases as the applied intensity decreases, which aligns with our expectations. Additionally, $\Delta T_{p\text{-}v}$ diminishes in magnitude as the linear transmittance of the finite aperture increases, eventually reaching a point where $\Delta T_{p\text{-}v}$ equals zero at $S = 1$. It is important to note that $S = 1$ represents the condition without an aperture, a setup typically used for investigating nonlinear absorption via the open z−scan technique. This suggests that selecting a smaller S value is the optimal condition for studying the phenomenon of nonlinear refraction. In our experiments, we conducted tests under three distinct applied intensities, each resulting in corresponding phase differences for a fixed $S = 0.01$. Figure 6 presents the magnitude of the peak-valley difference as a function of applied intensity (a), along with its fitting, and illustrates the nonlinear phase shift (b).

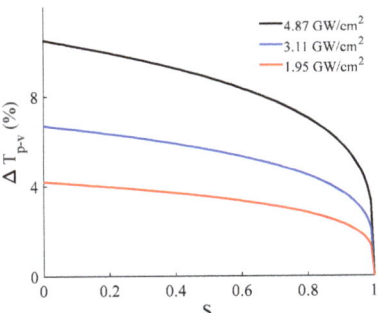

Figure 5. Difference between the normalized peak and valley transmittance as a function of linear transmittance of finite aperture (S) for different applied intensities.

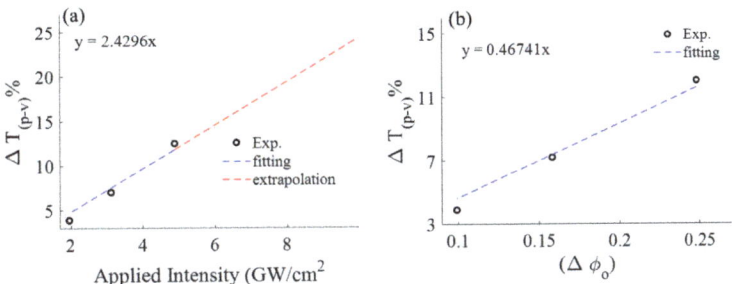

Figure 6. Percentage difference between the normalized peak and valley transmittance as a function of (**a**) applied intensity and (**b**) nonlinear phase shift ($\Delta\phi_o$) for linear transmittance of finite aperture (0.01).

Figure 6b displays the slope of $\Delta T_{p\text{-}v}$ as a function of the phase shift at the focus point. The estimation yields a linear coefficient of ~0.46 which aligns with z−scan model [16]. In addition, for a higher S, the linear coefficients (slopes) become smaller, which is expected in the z−scan as shown in Figure 7 [16].

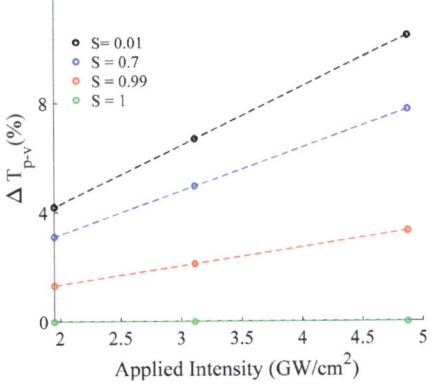

Figure 7. Percentage difference between the normalized peak and valley transmittance as a function of applied intensity for various linear transmittance of finite aperture (S).

The absolute value of third-order nonlinear susceptibility is calculated using

$$\left|\chi^{(3)}\right| = \sqrt{\left|\text{Re}\chi^{(3)}\right|^2 + \left|\text{Im}\chi^{(3)}\right|^2} \quad (4)$$

where $\text{Re}\chi^{(3)} = (4/3)n_o^2\varepsilon_o c\gamma$ and $\text{Im}\chi^{(3)} = (1/3\pi)n_o^2\varepsilon_o c\lambda\beta$ are real and imaginary components of third-order nonlinearity, n_o is the linear refractive index [42], ε_o is the dielectric constant of the vacuum, and c is the velocity of light. The third-order nonlinear susceptibility of the GO monolayer in DI water was estimated to be $|\chi^{(3)}| \sim 9.55 \times 10^{-17}$ m^2/V^2, which is close to those in the literature review ($\sim 8.97 \times 10^{-18}$ m^2/V^2 GO in DI water [27] and $\sim 5.12 \times 10^{-16}$ m^2/V^2 GO in ethanol [31]).

The rise time of a thermal lens in an aqueous liquid is determined using the acoustic transit time, denoted as $\tau = w_o/v_s$ [16] where $v_s \approx 1437$ m/s [45], is the velocity of sound in the water at room temperature. Consequently, we calculate τ to be approximately 10.1 ns, nearly double the pulse width of around 6 ns. Furthermore, it is worth noting that the thermal diffusion time in water exceeds the pulse width [46]. It implies that there is non-uniform heating induced by the 6 nm pulsed laser, leading to a nonlinear index change at the focus axis. Consequently, it suggests that the third-order nonlinear optical response encompasses both electronic effects (Kerr effect) and thermal effects contributing to γ. The validity of the nonlinear refraction coefficient's polarity is additionally confirmed through the measurement of transmittance as a function of applied intensity, employing the I−scan technique, as illustrated in Figure 8. The transmittance observed in the I−scan is contingent on multiple factors, including the applied intensity, nonlinear absorption, nonlinear refraction, sample position, Rayleigh range, effective sample length, and the distance between the sample and the aperture [16]. In the I−scan, the GO dispersion was positioned at the valley of the closed z−scan traces, which revealed a reduction in transmittance with increasing applied intensity. This phenomenon occurs because the transmittance beam experiences increased diffraction as the peak intensity rises. At the valley position, the intensity-dependent total absorption ($\alpha(I) = \alpha_o + \beta I$) and total refraction ($n(I) = n_o + \gamma I$) collectively contribute to the laser power attenuation capacity, signifying the potential of GO nanoflakes as valuable materials for optical power limiting. Such materials can play a critical role in safeguarding the eyes and sensors from high-power radiation.

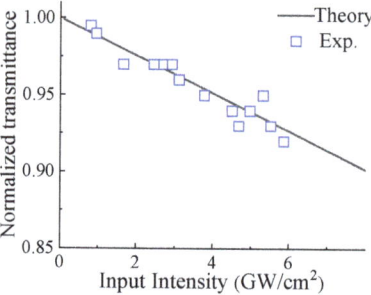

Figure 8. Normalized transmittance as a function of applied peak intensity at the valley position of valley-peak of closed z−scan traces.

4. Conclusions

The third-order optical nonlinear properties of GO were studied through the z−scan technique using a single wavelength at a 532 nm laser source. The positive nonlinear (reverse saturable) absorption and nonlinear (valley-peak) refraction properties were observed through the open z−scan and closed z−scan techniques, respectively. Given the linear absorption spectra below the 500 nm region, the 532 nm excitation source likely facilitated nonlinear excitation through either two-photon excitation or a two-step excitation process, leading to reverse saturable absorption. This finding indicates that graphene oxide possesses a larger excited-state absorption cross-section than its ground state, making it a promising candidate for optical limiters. The estimated coefficients for nonlinear absorption and nonlinear refraction were approximately ~2.62×10^{-8} m/W and 3.9×10^{-15} m^2/W, respectively. Moreover, we observed a reduction in laser transmittance through graphene oxide at higher intensities using an I−scan, suggesting its potential application in optical power-limiting devices in future technological advancements.

Author Contributions: Conceptualization, F.J.S., W.-J.K. and T.N.; methodology, T.N. and U.P.; formal analysis, T.N. and U.P.; resources, F.J.S.; writing—original draft preparation, T.N. and U.P.; writing—review and editing, F.J.S., B.T., T.N., U.P. and W.-J.K.; visualization, T.N.; supervision, F.J.S. and B.T.; funding acquisition, F.J.S. and W.-J.K. All authors have read and agreed to the published version of the manuscript.

Funding: This work was supported at Hampton by NASA NNX15AQ03A and ARO W911NF-15-1-0535; at UNC-Pembroke by 170040/368; and at K1 Solution R&D Center by the Korean NRF2020M3H4 A3081799.

Institutional Review Board Statement: Not applicable.

Informed Consent Statement: Not applicable.

Data Availability Statement: The data presented in this study are available on request from the corresponding author.

Conflicts of Interest: The authors declare no conflict of interest.

References

1. Geim, A.K.; Novoselov, K.S. The rise of graphene. *Nat. Mater.* **2007**, *6*, 183–191. [CrossRef] [PubMed]
2. Zhang, H.; Virally, S.; Bao, Q.; Ping, L.K.; Massar, S.; Godbout, N.; Kockaert, P. Z−scan measurement of the nonlinear refractive index of graphene. *Opt. Lett.* **2012**, *37*, 1856–1858. [CrossRef]
3. Bao, Q.; Zhang, H.; Wang, Y.; Ni, Z.; Yan, Y.; Shen, Z.; Loh, K.P.; Tang, D.Y. Atomic-Layer Graphene as a Saturable Absorber for Ultrafast Pulsed Lasers. *Adv. Funct. Mater.* **2009**, *19*, 3077–3083. [CrossRef]
4. Mizrahi, V.; Delong, K.W.; Stegeman, G.I.; Saifi, M.A.; Andrejco, M.J. Two-photon absorption as a limitation to all-optical Switching. *Opt. Lett.* **1982**, *14*, 1140–1142. [CrossRef] [PubMed]
5. Chantharasupawong, P.; Philip, R.; Narayanan, N.T.; Sudeep, P.M.; Mathkar, A.; Ajayan, P.M.; Thomas, J. Optical Power Limiting in Fluorinated Graphene Oxide: An Insight into the Nonlinear Optical Properties. *J. Phys. Chem. C* **2012**, *116*, 25955–25961. [CrossRef]
6. Siyar, M.; Maqsood, A.; Khan, S. Synthesis of mono layer graphene oxide from sonicated graphite flakes and their Hall effect measurements. *Mater. Sci. Pol.* **2014**, *32*, 292–296. [CrossRef]
7. Zhu, Y.; Murali, S.; Cai, W.; Li, X.; Suk, J.W.; Potts, J.R.; Ruoff, R.S. Graphene and Graphene Oxide: Synthesis, Properties, and Applications. *Adv. Mater.* **2010**, *22*, 3906–3924. [CrossRef]
8. Gupta, V.; Sharma, N.; Singh, U.; Arif, M.; Singh, A. Higher oxidation level in graphene oxide. *Optik* **2017**, *143*, 115–124. [CrossRef]
9. Ren, J.; Zheng, X.; Tian, Z.; Li, D.; Wang, P.; Jia, B. Giant third-order nonlinearity from low-loss electrochemical graphene oxide film with a high-power stability. *Appl. Phys. Lett.* **2016**, *109*, 221105. [CrossRef]
10. Loh, K.P.; Bao, Q.; Eda, G.; Chhowalla, M. Graphene oxide as a chemically tunable platform for optical applications. *Nat. Chem.* **2010**, *2*, 1015–1024. [CrossRef] [PubMed]
11. Jiang, X.F.; Polavarapu, L.; Neo, S.T.; Venkatesan, T.; Xu, Q.H. Graphene Oxides as Tunable Broadband Nonlinear Optical Materials for Femtosecond Laser Pulses. *J. Phys. Chem. Lett.* **2012**, *3*, 785–790. [CrossRef] [PubMed]
12. Xu, X.; Zheng, X.; He, F.; Wang, Z.; Subbaraman, H.; Wang, Y.; Jia, B.; Chen, R.T. Observation of Third-order Nonlinearities in Graphene Oxide Film at Telecommunication Wavelengths. *Sci. Rep.* **2017**, *7*, 9646. [CrossRef]
13. Burkins, P.; Kuis, R.; Basaldua, I.; Johnson, A.M.; Swaminathan, S.R.; Zhang, D.; Trivedi, S. Thermally managed Z−scan methods investigation of the size-dependent nonlinearity of graphene oxide in various solvents. *JOSAB* **2016**, *33*, 2395–2401. [CrossRef]
14. Krishna, M.B.M.; Venkatramaiah, N.; Venkatesan, R.; Rao, D.N. Synthesis and structural, spectroscopic and nonlinear optical measurements of graphene oxide and its composites with metal and metal free porphyrins. *J. Mater. Chem.* **2010**, *22*, 3059. [CrossRef]
15. Shan, Y.; Tang, J.; Wu, L.; Lu, S.; Dai, X.; Xiang, Y. Spatial self-phase modulation and all-optical switching of graphene oxide dispersions. *J. Alloys Compd.* **2019**, *771*, 900–904. [CrossRef]
16. Bahae, M.S.; Said, A.A.; Wei, T.H.; Hagan, D.J.; Stryland, E.W. Sensitive measurement of optical nonlinearities using a single beam. *IEEE J. Quantum Electron.* **1990**, *26*, 760–769. [CrossRef]
17. Durbin, S.D.; Arakelian, S.M.; Shen, Y.R. Laser-induced diffraction rings from a nematic-liquid-crystal film. *Opt. Lett.* **1981**, *6*, 411–413. [CrossRef] [PubMed]

18. Callen, W.R.; Huth, B.G.; Pantell, R.H. Optical Patterns of Thermally Self-Defocused Light. *Appl. Phys. Lett.* **1967**, *11*, 103–105. [CrossRef]
19. Kean, P.N.; Smith, K.; Sibbett, W. Spectral and temporal investigation of self-phase modulation and stimulated Raman scattering in a single-mode optical fibre. *IEE Proceeding* **1987**, *134*, 163–167. [CrossRef]
20. Neupane, T.; Wang, H.; Yu, W.W.; Tabibi, B.; Seo, F.J. Second order hyperpolarizability and all-optical-switching of intensity-modulated spatial self-phase modulation in $CsPbBr_{1.5}I_{1.5}$ perovskite quantum dot. *Opt. Laser Technol.* **2021**, *140*, 107090. [CrossRef]
21. Neupane, T.; Tabibi, B.; Kim, W.J.; Seo, F.J. Spatial Self-Phase Modulation in Graphene-Oxide Monolayer. *Crystals* **2023**, *13*, 271. [CrossRef]
22. Neupane, T.; Tabibi, B.; Seo, F.J. Spatial Self-Phase Modulation in Ws_2 and MoS_2 atomic layer. *Opt. Mater. Express* **2020**, *8*, 831–842. [CrossRef]
23. Parra, I.; Valbuena, S.; Racedo, S. Measurement of non-linear optical properties in graphene oxide using the Z−scan technique. *Spectrochim. Acta Part A Mol. Biomol. Spectrosc.* **2021**, *244*, 118833. [CrossRef]
24. Woodward, R.I.; Kelleher, E.J.R.; Howe, R.C.T.; Hu, G.; Torrisi, F.; Hasan, T.; Popov, S.V.; Taylor, J.R. Tunable Q-switched fiber laser based on saturable edge-state absorption in few-layer molybdenum disulfide (MoS_2). *Opt. Express* **2014**, *22*, 31113–31122. [CrossRef] [PubMed]
25. Wang, J.; Gu, B.; Wang, H.T.; Ni, X.W. Z−scan analytical theory for material with saturable absorption and two photon absorption. *Opt. Commun.* **2010**, *283*, 3525–3528. [CrossRef]
26. Bahae, M.S.; Said, A.A.; Stryland, E.W. High-sensitivity, single-beam n_2 measurements. *Opt. Lett.* **1989**, *14*, 955–957. [CrossRef] [PubMed]
27. Kang, L.; Sato, R.; Zhang, B.; Takeda, Y.; Tang, J. Experimental dispersion of the third-order optical susceptibility of graphene oxide. *Opt. Mater. Express* **2020**, *10*, 3041–3050. [CrossRef]
28. Chantharasupawong, P.; Philip, R.; Endo, T.; Thomas, J. Enhanced optical limiting in nanosized mixed zinc ferrites. *Appl. Phys. Lett.* **2012**, *100*, 221108. [CrossRef]
29. Philip, R.; Chantharasupawong, P.; Qian, H.; Jin, R.; Thomas, J. Evolution of nonlinear optical properties: From gold atomic clusters to plasmonic nanocrystals. *Nano Lett.* **2012**, *12*, 4661–4667. [CrossRef]
30. Seo, J.T.; Ma, S.M.; Yang, Q.; Creekmore, L.; Battle, R.; Brown, H.; Jackson, A.; Skyles, T.; Tabibi, B.; Yu, W.; et al. Large resonant third-order optical nonlinearity of CdSe nanocrystal quantum dots. *J. Phys. Conf. Ser.* **2006**, *38*, 91–95. [CrossRef]
31. Liaros, N.; Iliopoulos, K.; Stylianakis, M.M.; Koudoumas, E.; Couris, S. Optical limiting action of few layered graphene oxide dispersed in different solvents. *Opt. Mater.* **2013**, *36*, 112–117. [CrossRef]
32. Kumara, K.; Kindalkar, V.S.; Serrao, F.J.; Shetty, T.C.S.; Patil, P.S.; Dharmaprakash, S.M. Enhanced nonlinear optical absorption in defect enriched graphene oxide and reduced graphene oxide using continuous wave laser z−scan technique. *Mater. Today Proc.* **2022**, *55*, 186–193. [CrossRef]
33. Muruganandi, G.; Saravanan, M.; Vinitha, G.; Jessie Raj, M.B.; Girisun, T.C.S. Effect of reducing agents in tuning the third-order optical nonlinearity and optical limiting behavior of reduced graphene oxide. *Chem. Phys.* **2017**, *488*, 55–61. [CrossRef]
34. Naghani, M.E.; Neghabi, M.; Zadsar, M.; Ahangar, H.A. Synthesis and characterization of linear/nonlinear optical properties of graphene oxide and reduced graphene oxide-based zinc oxide nanocomposite. *Sci. Rep.* **2023**, *13*, 1496. [CrossRef] [PubMed]
35. Feng, M.; Zhan, H.; Chen, Y. Nonlinear optical and optical limiting properties of graphene families. *Appl. Phys. Lett.* **2010**, *96*, 033107. [CrossRef]
36. Fang, C.; Dai, B.; Hong, R.; Tao, C.; Wang, Q.; Wang, X.; Zhang, D.; Zhuang, S. Tunable optical limiting optofluidic device filled with graphene oxide dispersion in ethanol. *Sci. Rep.* **2015**, *5*, 15362. [CrossRef]
37. Ebrahimi, M.; Zakery, A.; Karimipour, M.; Molaei, M. Nonlinear optical properties and optical limiting measurements of graphene oxide—$Ag@TiO_2$ compounds. *Opt. Mater.* **2016**, *57*, 146–152. [CrossRef]
38. Shi, H.; Wang, C.; Sun, Z.; Zhou, Y.; Jin, K.; Redfern, S.A.T.; Yang, G. Tuning the nonlinear optical absorption of reduced graphene oxide by chemical reduction. *Opt. Express* **2014**, *22*, 19375–19385. [CrossRef] [PubMed]
39. Silva, T.P.S.; Silva, A.A.; Oliveira, M.C.D.; Souza, P.R.; Filho, E.C.S.; Garcia, H.A.; Costa, J.C.S.; Santos, F.E.P. Biosynthesis of Ag@Au bimetallic nanoparticles from *Hymenaea courbaril* extract (Jatobá) and nonlinear optics properties. *J. Mol. Liq.* **2023**, *389*, 122641. [CrossRef]
40. Coleman, J.N.; Khan, U.; Young, K.; Gaucher, A.; De, S.; Smith, R.J.; Shvets, I.V.; Arora, S.K.; Stanton, G.; Kim, H.; et al. Two-Dimensional Nanosheets Produced by Liquid Exfoliation of Layered Materials. *Science* **2011**, *331*, 568–571. [CrossRef]
41. Neupane, T.; Yu, S.; Rice, Q.; Tabibi, B.; Seo, F.J. Third-order optical nonlinearity of tungsten disulfide atomic layer with resonant excitation. *Opt. Mat.* **2019**, *96*, 109271. [CrossRef]
42. Kimiagar, S.; Abrinaei, F. Effect of temperature on the structural, linear, and nonlinear optical properties of MgO-doped graphene oxide nanocomposites. *Nanophotonics* **2017**, *7*, 243. [CrossRef]
43. Neupane, T.; Rice, Q.; Jung, S.; Tabibi, B.; Seo, F.J. Cubic Nonlinearity of Molybdenum Disulfide Nanoflakes. *J. Nanosci. Nanotechnol.* **2020**, *20*, 4373–4375. [CrossRef] [PubMed]
44. Kwak, C.H.; Lee, Y.L.; Kim, S.G. Analysis of asymmetric Z−scan measurement for large optical nonlinearities in an amorphous As_2S_3 thin film. *J. Opt. Soc. Am. B* **1999**, *16*, 600–604. [CrossRef]

45. Martin, K.; Spinks, D. Measurement of the speed of sound in ethanol/water mixture. *Ultrasound Med. Biol.* **2001**, *27*, 289–291. [CrossRef]
46. Khabibullin, V.R.; Usoltseva, L.O.; Galkina, P.A.; Galimova, V.R.; Volkov, D.S.; Mikheev, I.V.; Proskurnin, M.A. Measurement Precision and Thermal and Absorption Properties of Nanostructure in Aqueous Solutions by Transient and Steady-State Thermal-Lens Spectrometry. *Phys. Chem.* **2023**, *3*, 156–197. [CrossRef]

Disclaimer/Publisher's Note: The statements, opinions and data contained in all publications are solely those of the individual author(s) and contributor(s) and not of MDPI and/or the editor(s). MDPI and/or the editor(s) disclaim responsibility for any injury to people or property resulting from any ideas, methods, instructions or products referred to in the content.

Systematic Review

Application of Graphene Oxide in Oral Surgery: A Systematic Review

Francesco Inchingolo [1,†], Angelo Michele Inchingolo [1,†], Giulia Latini [1], Giulia Palmieri [1], Chiara Di Pede [1], Irma Trilli [1], Laura Ferrante [1], Alessio Danilo Inchingolo [1,*], Andrea Palermo [2], Felice Lorusso [3], Antonio Scarano [3] and Gianna Dipalma [1,†]

[1] Interdisciplinary Department of Medicine, University of Bari "Aldo Moro", 70124 Bari, Italy; francesco.inchingolo@uniba.it (F.I.); angeloinchingolo@gmail.com (A.M.I.); dr.giulialatini@gmail.com (G.L.); giuliapalmieri13@gmail.com (G.P.); c.dipede1@studenti.uniba.it (C.D.P.); trilliirma@gmail.com (I.T.); lauraferrante79@virgilio.it (L.F.); giannadipalma@tiscali.it (G.D.)
[2] College of Medicine and Dentistry, Birmingham B4 6BN, UK; andrea.palermo2004@libero.it
[3] Department of Innovative Technologies in Medicine and Dentistry, University of Chieti–Pescara, 66100 Chieti, Italy; drlorussofelice@gmail.com (F.L.); ascarano@unich.it (A.S.)
* Correspondence: ad.inchingolo@libero.it; Tel.: +39-3345337663
† These authors contributed equally to this work.

Abstract: The current review aims to provide an overview of the most recent research in the last 10 years on the potentials of graphene in the dental surgery field, focusing on the potential of graphene oxide (GO) applied to implant surfaces and prosthetic abutment surfaces, as well as to the membranes and scaffolds used in Guided Bone Regeneration (GBR) procedures. "Graphene oxide" and "dental surgery" and "dentistry" were the search terms utilized on the databases Scopus, Web of Science, and Pubmed, with the Boolean operator "AND" and "OR". Reviewers worked in pairs to select studies based on specific inclusion and exclusion criteria. They included animal studies, clinical studies, or case reports, and in vitro and in vivo studies. However, they excluded systematic reviews, narrative reviews, and meta-analyses. Results: Of these 293 studies, 19 publications were included in this review. The field of graphene-based engineered nanomaterials in dentistry is expanding. Aside from its superior mechanical properties, electrical conductivity, and thermal stability, graphene and its derivatives may be functionalized with a variety of bioactive compounds, allowing them to be introduced into and improved upon various scaffolds used in regenerative dentistry. This review presents state-of-the-art graphene-based dental surgery applications. Even if further studies and investigations are still needed, the GO coating could improve clinical results in the examined dental surgery fields. Better osseointegration, as well as increased antibacterial and cytocompatible qualities, can benefit GO-coated implant surgery. On bacterially contaminated implant abutment surfaces, the CO coating may provide the optimum prospects for soft tissue sealing to occur. GBR proves to be a safe and stable material, improving both bone regeneration when using GO-enhanced graft materials as well as biocompatibility and mechanical properties of GO-incorporated membranes.

Keywords: graphene oxide (GO); dental surgery; graphene coating; oxide materials

1. Introduction

The landscape of modern healthcare has been continuously shaped by groundbreaking advancements in materials science and technology [1]. One such revolutionary material that has garnered significant attention is graphene, a two-dimensional carbon allotrope characterized by its exceptional properties [2–4]. It is represented by a hexagonal honeycomb structure in which each atom is able to bond with three adjacent atoms (Figure 1) [5–7]. Graphene's unrivaled mechanical strength, electrical conductivity, and biocompatibility have sparked interest across various scientific domains. In the realm of oral surgery, where precision, efficacy, and patient well-being are paramount, the integration of graphene holds tremendous promise for pushing the boundaries of traditional approaches and ushering in a new era of surgical innovation [8,9].

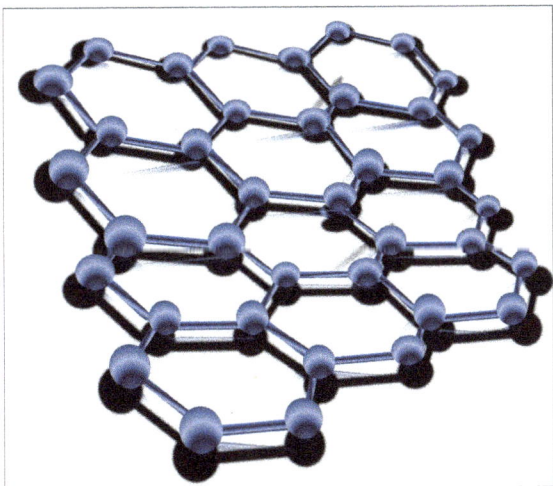

Figure 1. Crystal lattice of graphene with a hexagonal honeycomb structure in which each atom is able to bond with three adjacent atoms.

Promising uses for graphene include lots of biomedical fields [10]. Graphene's electrical conductivity, high specific surface area, and mechanical robustness can help transdermal biosensors provide signals with greater precision and repeatability to monitoring molecules and biomarkers [11]. Graphene has several potential uses as a consequence, including drug delivery systems, and has attracted a lot of interest in the field of biomedical 3D printing [10,12]. It has recently been demonstrated his impact in neurotherapeutics for neuroimaging, neuro-oncology, and neuro-surgery [13].

Oral surgery encompasses a spectrum of procedures, ranging from routine tooth extractions to complex maxillofacial reconstructions [8,14–17]. The quest for enhanced patient outcomes and the refinement of surgical techniques has been an ongoing pursuit in this field [18]. Graphene, with its unique attributes, emerges as a material with the potential to redefine the landscape of oral surgery, offering novel solutions for challenges that have long persisted and introducing avenues for previously unexplored possibilities [19].

At the heart of graphene's allure lies its remarkable physical properties. Structurally, graphene consists of a single layer of carbon atoms arranged in a hexagonal lattice [20]. This arrangement imparts extraordinary mechanical strength, rendering graphene the strongest material ever tested. Such mechanical robustness holds promise for oral surgery, where materials capable of withstanding physiological forces while promoting integration with surrounding tissues are highly sought after [21]. Graphene's strength can be harnessed in the fabrication of dental implants, orthodontic devices, and reconstructive scaffolds that maintain structural integrity and support tissue regeneration [22].

Graphene's electrical conductivity is equally intriguing. Its high electron mobility opens the door to applications involving electrical stimulation and biosensing. In the context of oral surgery, this property could lead to the development of implantable devices capable of monitoring healing processes in real time, thereby enabling timely interventions in case of complications [23]. Additionally, the integration of graphene-based sensors could enhance the accuracy of surgical procedures, offering surgeons immediate feedback and aiding in precise tissue manipulation [24].

A cornerstone of successful oral surgery is the interaction between surgical materials and the complex biological environment [25]. In-depth research has been conducted on the biocompatibility of graphene, and studies have shown that it may be able to facilitate cellular adhesion, proliferation, and differentiation [26]. The promise of graphene as a scaffold material for tissue engineering applications in oral surgery is highlighted by this feature. Graphene scaffolds have the potential to speed up the creation of bioengineered oral tissues, bone regeneration, and wound healing by creating an environment that is favorable for cellular growth [27].

Moreover, graphene's interactions with immune cells have raised intriguing possibilities for modulating immune responses during surgical interventions [28,29]. This presents the potential to reduce inflammation, enhance tissue integration, and ultimately improve patient recovery and comfort post-surgery [28].

The prevention of post-operative infections remains a critical challenge in oral surgery [30]. Graphene's inherent antibacterial properties have attracted considerable attention. Its unique interaction with bacterial cell membranes disrupts their structural integrity, rendering them susceptible to elimination [31,32]. This property could be leveraged to develop antimicrobial coatings for surgical instruments, implants, and wound dressings [33–35]. By mitigating bacterial colonization and biofilm formation, graphene-based materials could substantially reduce the risk of infections, leading to improved patient outcomes and decreased reliance on antibiotics [36,37].

While the potential applications of graphene in oral surgery are undeniably exciting, several challenges must be addressed to ensure safe and effective clinical implementation. The scalable synthesis of graphene materials suitable for surgical use, long-term biocompatibility assessments, and regulatory approvals are among the foremost challenges. Additionally, the development of standardized surgical protocols and techniques for incorporating graphene into existing procedures is essential to ensure seamless integration and optimal outcomes [23,38].

As the frontiers of materials science and oral surgery intersect, graphene emerges as a transformative force poised to reshape the landscape of surgical practices [34]. Its extraordinary mechanical, electrical, and biocompatible properties offer novel solutions to age-old challenges while presenting unprecedented opportunities for innovation [39,40]. This comprehensive exploration sheds light on the manifold applications of graphene in oral surgery, emphasizing the potential to revolutionize patient care, surgical techniques, and the overall trajectory of the field [41,42]. As researchers and clinicians continue to unravel graphene's potential, the future of oral surgery appears brighter and more promising than ever before [43,44].

2. Materials and Methods

2.1. Protocol and Registration

This systematic review was conducted by the standards of the Preferred Reporting Items for Systematic Reviews and Meta-analysis (PRISMA), and it was submitted to PROSPERO with number ID 453609.

2.2. Search Processing

Graphene oxide, dental surgery, and dentistry were the search terms utilized on the databases (Scopus, Web of Science, and Pubmed) to select the papers under evaluation, with the Boolean operators "AND" and "OR".

The search was restricted to just items released in English during the previous ten years (July 2013–July 2023).

2.3. Eligibility Criteria

The reviewers, who worked in pairs, chose works that satisfied the following criteria for inclusion: (1) animal studies; (2) clinical studies or case reports; and (3) in vitro and in vivo studies.

Exclusion criteria were systematic reviews, narrative reviews, and meta-analyses.

2.4. Data Processing

The screening procedure, which was carried out by reading the article titles and abstracts chosen in the earlier identification step, allowed for the exclusion of any publications that varied from the themes looked at.

The complete text of publications that had been determined to match the predetermined inclusion criteria was then read.

Reviewer disagreements on the choice of the article were discussed and settled.

2.5. Quality Assessment

The quality of the included papers was assessed by two reviewers, RF and EI, using the reputable Cochrane risk-of-bias assessment for randomized trials (RoB 2). The following six areas of possible bias are evaluated by this tool: random sequence generation, allocation concealment, participant and staff blinding, outcome assessment blinding, inadequate outcome data, and selective reporting. A third reviewer (FI) was consulted in the event of a disagreement until an agreement was reached.

3. Results

Keyword searches of the Web of Science (55), Scopus (38), and Pubmed (200) databases yielded a total of 293 articles.

The subsequent elimination of duplicates (61) resulted in the inclusion of 232 articles.

Of these 232 studies, 213 were excluded because they deviated from the previously defined inclusion criteria.

The screening phase ended with selecting 19 publications for this work.

The results of each study are reported in Figure 2.

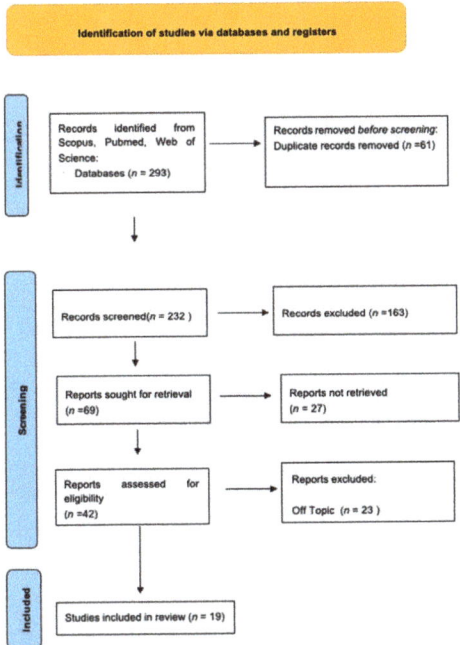

Figure 2. PRISMA flowchart.

The study data was selected by analyzing the study design, number of patients, intervention, and outcomes (Table 1).

Table 1. Characteristics of the studies included in the analysis.

Authors (Year)	Type of the Study	Aim of the Study	Materials	Results
Eshghinejad et al. [45] 2019	In vitro study	This article details our research into the electrophoretic deposition of composite materials consisting of BG-GO onto titanium alloy implants, aiming to enhance their antibacterial capabilities and biocompatibility.	Comparison of samples coated with BG-GO versus BG alone.	Enhanced antibacterial performance was observed in BG-GO-coated samples compared to BG-only coatings, with improved effectiveness as GO content increased. The BG-GO composite coating demonstrated favorable biocompatibility based on cell adhesion tests, indicating that the presence of GO did not hinder cell attachment to the alloy surface. Consequently, the BG-GO composite coatings, fabricated using the EPD technique and exhibiting these attributes, hold significant promise as a viable option for bone implant applications.
Ren et al. [46] 2019	In vitro study	The aim is to create a drug delivery system by coating titanium foils with graphene oxide and titanate, with the goal of boosting the growth and differentiation of rBMSCs towards osteogenesis.	GO sheets, generated using a modified Hummer's method, were integrated with bioactive titanate onto titanium implants (referred to as GO-Ti) prior to reduction (resulting in rGO-Ti). The growth of rBMSCs on these surfaces was evaluated through mRNA expression and alkaline phosphatase activity.	The findings demonstrated excellent performance of the Dexamethasone-loaded surface (DEX-GO-Ti) in promoting cell proliferation. On DEX-GO-Ti, significant expression of osteogenic differentiation-related proteins, mRNA, and calcium was observed in RMBSCs.
Park et al. [47] 2023	In vitro study	Atmospheric pressure plasma was employed to apply a coating of graphene possessing photothermal characteristics onto a zirconia surface.	Utilizing an atmospheric pressure plasma generator (PGS-300, Expantech, Suwon, Republic of Korea), an Ar/CH4 gas combination was applied to a zirconia sample at a power level of 240 W and a flow rate of 10 L/min.	The category where the zirconia sample, covered with graphene oxide, underwent near-infrared ray exposure and exhibited a noteworthy decrease in the attachment of *S. mutans* and *P. gingivalis* in comparison to the non-irradiated group.
Cheng et al. [48] 2022	In vitro study	The aim of the study was to evaluate the antibacterial properties and cytocompatibility of a novel composite coating containing GO and the antimicrobial peptide (AMP) Nal-P-113 on a smooth titanium surface.	Smooth titanium surface coated with GO and antimicrobial peptide (AMP) Nal-P-113.	The Nal-P-113-loaded GO coating exhibited potent antibacterial activity against both Gram-positive (*S. mutans*) and Gram-negative (*P. gingivalis*) bacteria while maintaining biocompatibility with HGF cells.

Table 1. Cont.

Authors (Year)	Type of the Study	Aim of the Study	Materials	Results
Jang et al. [49] 2021	In vitro study	To investigate how the application of GO onto a zirconia surface influences the attachment of bacteria and the activation of osteoblasts.	The atmospheric pressure plasma generator (PGS-300) was used to apply a blend of Ar/CH4 gas onto zirconia samples, dividing them into two groups: uncoated (Zr group) and graphene oxide-coated (Zr-GO group).	GO-coated zirconia effectively obstructs S. mutans bacteria adhesion, promoting osteoblast growth and specialization. This suggests its potential in combating peri-implantitis by deterring bacterial attachment and enhancing bone adhesion, thereby improving implant success rates.
Guo et al. [50] 2021	In vitro study	To test the antimicrobial effects of PEEK-PDA-GO surfaces	Antibacterial and cellular tests	PEEK-PDA-GO effectively inhibits microorganisms such as Streptococcus mutans, Fusobacterium nucleatum, and Porphyromonas gingivalis, promoting strong human gingival fibroblast adherence and proliferation.
Qin et al. [51] 2021	In vitro study	To test the effect of GO-carbon fibers (CF)-PEE coating on Titanium implants	Physiochemical and cellular tests	GO-CF-PEEK: - Antimicrobial effects. - Reducing the coefficient of friction and improving wear resistance. - Cytocompatibility on murine fibroblasts.
Qin et al. [52] 2021	In vitro and in vivo study	To test biological safety and osseointegration of GO-CF-PEEK coatings.	Cellular tests and in vivo analysis of osseointegration.	GO-CF-PEEK:Surface hydrophilicity was increased. Porous nanostructures improved early cell activities and osseointegration.
Qin et al. [53] 2020	In vitro study	To determine whether polymicrobial biofilms can be removed using GO.	The study examined in vitro biofilm formation on titanium surfaces using brushing alone, varying GO concentrations, combined treatments, and no therapy.	GO at high concentrations removed bacteria and prevented biofilm reformation in combination with brushing (Group GB). The BMSCs' osteogenic capacity was increased on the GO Ti surfaces.
Patil et al. [54] 2020	In vitro study	To determine the effects of titanium alloy, graphene, and reduced graphene oxide (rGO) on tension and distortion at the implant.	Finite element analysis (FEA).	Titanium implants had better mechanical behavior than graphene when coated with rGO.
Jeong-Woo Kim et al. (2017) [55]	In vivo	To evaluate the effect of biphasic calcium phosphate (BCP) coated with reduced graphene oxide (rGO) as bone graft materials on bone regeneration.	- BCP coated with rGO fabricated at various concentrations. - Cell viability tests conducted at different rGO concentrations.	- New bone formation evaluated using micro-CT and histological analysis. - Effectiveness of rGO-coated BCP on osteogenesis. - Importance of composite concentration.

Table 1. Cont.

Authors (Year)	Type of the Study	Aim of the Study	Materials	Results
Erika Nishida et al. (2016) [56]	In vitro	To ascertain whether the graphene oxide scaffold promoted bone induction in the extractive alveoli of dog teeth.	- Fabrication of GO-applied scaffold and dispersion on collagen sponge scaffold - Characterization using SEM, physical testing, cell seeding, and rat subcutaneous implant testing - Implantation of GO scaffold into dog tooth extraction socket - Histological observations at 2 weeks post-surgery.	- Improved physical strength, enzyme resistance, calcium, and protein adsorption due to GO application. - Increased osteoblastic cell proliferation with GO application. - Good biocompatibility observed through rat subcutaneous tissue response. - Enhanced bone formation in dogs.
Jong Ho Lee et al. (2015) [57]	In vitro	To examine whether reduced graphene oxide (rGO) and hydroxyapatite (HAp) nanocomposites (rGO/HAp NC) could enhance MC3T3-E1 preosteoblast osteogenesis and promote new bone formation.	- Examination of the potential of graphene-based hybrid composites for cellular differentiation and tissue regeneration.	- Synergistic promotion of osteodifferentiation without hindering proliferation observed with rGO/HAp combination. - Graphene-based composites found to have osteogenesis stimulation potential.
Izumi Kanayama et al. (2014) [58]	In vitro	To examine the bioactivity of graphene oxide (GO) and Reduced graphene oxide (RGO) films and collagen scaffolds coated with GO and RGO.	- Evaluation of GO and RGO films' bioactivity and collagen scaffolds coated with GO - Biological properties assessed using SEM, atomic force microscopy, calcium adsorption tests, and MC3T3-E1 cell seeding. - Implantation of scaffolds into rat subcutaneous tissue, followed by DNA content and cell ingrowth measurements 10 days post-surgery.	- GO and RGO films possess distinct biological properties: enhanced calcium adsorption and alkaline phosphatase activity, promoting osteogenic differentiation; - GO- and RGO-coated scaffolds exhibit higher compressive strengths compared to non-coated scaffolds. - RGO-coated scaffolds are more bioactive than GO-coated scaffolds.
Chingis Daulbayev et al. (2022) [59]	In vitro	The GO/HAp composite prepared was dispersed in biodegradable polymer-polycaprolactone (PCL) in order to design a composite scaffold with the aim of enhancing osteogenic differentiation of osteoblasts for potential medical application	- Utilization of biodegradable polycaprolactone (PCL), graphene oxide (GO), and calcium hydroxyapatite (HAp). - Dispersal of GO/HAp composite in PCL for composite scaffold creation aimed at enhancing osteogenic differentiation of osteoblasts.	- Obtained composite scaffold suitable for bone tissue regeneration with antimicrobial properties.

Table 1. *Cont.*

Authors (Year)	Type of the Study	Aim of the Study	Materials	Results
Milena Radunovic et al. (2017) [60]	In vitro	To investigate the biocompatibility of GO-coated collagen membranes on human dental pulp stem cells (DPSCs), focusing on the cytotoxicity of biomaterials and the ability to promote the differentiation process of DPSCs and to control the induction of the inflammatory event.	Investigation of biocompatibility of GO-coated collagen membranes on human dental pulp stem cells (DPSCs).	- Faster DPSCs differentiation into odontoblasts/osteoblasts induced by GO-coated membranes - Potential as an alternative to conventional membranes for efficient bone formation and improved clinical performance.
Letizia Ferroni et al. (2022) [61]	In vitro	The amount of Rgo filler was defined to achieve a biocompatible and antibacterial PCL-based surface that supports human mesenchymal stem cell (MSC) adhesion and differentiation. Compounds containing three different percentages of Rgo were tested.	Evaluation of PCL-based surfaces with reduced graphene oxide (Rgo) nanofillers for bone regeneration in dentistry.	- Different percentages of rGO filler in PCL tested for biocompatibility and antibacterial properties - All scaffolds exhibit biocompatibility, antibacterial properties, adhesion, and differentiation of human mesenchymal stem cells (MSCs).
Elham-Sadat Motiee et al. (2023) [62]	In vitro	Poly-3 hydroxybutyrate-chitosan (PC) scaffolds reinforced with graphene oxide (GO) were fabricated by the electrospinning method to evaluate the possible increase in the biomechanical properties of the scaffolds.	- Development of Poly-3 hydroxybutyrate-chitosan (PC) scaffolds reinforced with graphene oxide (GO) via electrospinning method - Investigation of how GO reinforcement affects fibers diameter, thermal capacity, surface hydrophilicity, mechanical properties, and degradation of the nanocomposite scaffolds.	Improved physicochemical, mechanical, and biological properties demonstrate the potential of PCG nanocomposite scaffolds for bone tissue engineering.
Alana P C Souza et al. (2022) [63]	In vitro	Develop a chitosan-xanthan (CX) membrane associated with hydroxyapatite (HA) and different concentrations of graphene oxide (GO).	The study developed a chitosan-xanthan membrane with HA and GO concentrations, characterized using various techniques, including X-ray diffraction, FTIR, Raman spectroscopy, SEM, contact angle, tensile strength, bioactivity, and cell viability.	- Membranes with non-porous, homogeneous surfaces, hydrophilic nature, higher tensile strength, and reliability for guided bone regeneration therapies are essential for various applications.

4. Discussion

4.1. Implant and Abutment surfaces

This review discusses the potential of graphene oxide (GO) as a promising nanomaterial with exceptional physical and chemical properties. Recent research has focused on its applications in biomedical fields such as tissue engineering, antimicrobial materials, and implants [64]. The review examines the use of graphene to functionalize dental implant surfaces and its interactions with host tissue. Graphene is a single layer of sp2 hybridized carbon atoms arranged in a honeycomb lattice, known for its remarkable strength, elasticity, and electrical characteristics [34,65]. GO and reduced graphene oxide (rGO) are its primary derivatives. Due to their biocompatibility, low toxicity, hydro-solubility, and reactive oxygen groups, studies suggest that graphene and GO can support tissue regeneration, cell differentiation, and proliferation [1–3]. They also enhance the bioactivity and mechanical performance of biomaterials and can serve as carriers for drugs and biomolecules [6,54,65–69].

In order to manage infection and stop bone loss, peri-implantitis therapy must remove polymicrobial biofilms from the implant site and lessen tissue invasion. Brushing and GO at high concentrations effectively decontaminate biofilms from exposed titanium surfaces, as shown by Qin et al. [53].

The study by Ren et al. utilized GO as a coating on titanium foils to deliver drugs and enhance the growth and differentiation of rat bone mesenchymal stem cells (rBMSCs) into osteoblasts (Figure 3) [46]. The researchers incorporated dexamethasone (DEX) onto GO-coated titanium implants, resulting in improved absorption and sustained release of the drug. The DEX-GO-Ti substrates showed higher rBMSC proliferation compared to control and DEX-rGO-coated substrates [46]. Moreover, rBMSCs exhibited enhanced osteogenic differentiation on DEX-GO-Ti and DEX-rGO-Ti surfaces. This approach effectively controlled the bioactivity of titanium implants, showing promise for advancements in dentistry applications [46].

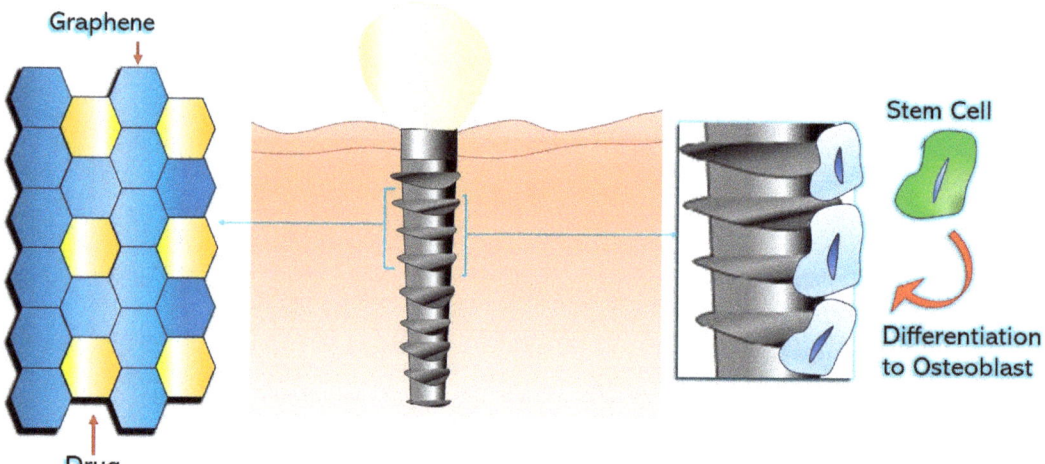

Figure 3. GO-coated implant surface enhances the differentiation of bone mesenchymal stem cells.

This study focuses on enhancing the antibacterial and cytocompatible attributes of titanium alloy implants by employing electrophoretic deposition to create bioglass (BG)-GO composites [45]. The resultant BG-GO composites formed a consistent and dense coating layer, measuring 50–55 μm in thickness [45]. This coating displayed enhanced resistance against corrosion and heightened antibacterial efficacy in comparison to samples coated solely with BG [45]. This antibacterial effectiveness escalated with an increase in

GO content. Cell adhesion findings indicated favorable biocompatibility of the BG-GO composite coating. Furthermore, the inclusion of GO in the BG-GO coating did not impede cell attachment to the alloy sample [45]. Consequently, the electrophoretic deposition method for creating BG-GO composite coatings with these beneficial traits presents a promising alternative for bone implant applications [45].

The physiochemical properties of GO-carbon fibers (CF)-PEEK on titanium implants were analyzed by Qin et al., who revealed that these coatings might greatly reduce the coefficient of friction of alloy and improve wear resistance [51]. In addition, GO-CF-PEEK showed biological safety and improved osteointegration [52].

Photothermal therapy (PTT), an alternative antibacterial treatment, has a substantial impact on deactivating oral microbiota. The study by Park et al. involved applying graphene possessing photothermal characteristics onto a zirconia surface using atmospheric pressure plasma, followed by an assessment of its antibacterial effects against oral bacteria [47]. To coat graphene oxide onto zirconia specimens, an atmospheric pressure plasma generator (PGS-300, Expantech, Suwon, Republic of Korea) was employed, utilizing an Ar/CH4 gas mixture at a power of 240 W and a flow rate of 10 L/min [47]. Notably, the group subjected to near-infrared irradiation after coating the zirconia specimen with graphene oxide exhibited a significant decrease in *S. mutans* and *P. gingivalis* adhesion compared to the non-irradiated group [47]. This reduction in oral microbiota activity was attributed to the photothermal effect on the zirconia surface coated with graphene oxide, which demonstrated photothermal properties [47].

Cheng et al. analyzed a new type of coating using a combination of GO and the antimicrobial peptide (AMP) Nal-P-113 on a smooth titanium surface [48]. The study evaluates the effectiveness of this coating at fighting bacteria and whether it is compatible with cells. The findings revealed that Nal-P-113 was gradually released from the composite coating over time when tested in a lab setting [48]. The GO coating loaded with Nal-P-113 demonstrated strong antibacterial properties against both Streptococcus mutans and Porphyromonas gingivalis, with no noticeable harm to human gingival fibroblast (HGF) cells [48]. However, further refinement is necessary to optimize the Nal-P-113-loaded GO coating for its potential to prevent infection and promote healing in the tissues surrounding implants [48]. The same results were obtained by Guo et al., who analyzed GO-modified Polyetheretherketone (PEEK) implant abutment surfaces grafted with dopamine [50].

The study by Jang et al. aimed to investigate the impact of applying GO onto a zirconia (Zr) surface on bacterial bonding and osteoblast activation [49]. Two groups of zirconia samples were compared: one without coating (Zr control) and another with GO coating (Zr-GO) [49]. Analysis through a scanning electron microscope confirmed successful GO deposition on the Zr-GO group. *S. mutans* bacterial attachment and growth were significantly reduced on the Zr-GO surface, while the attachment of MC3T3-1 cells remained similar, but their growth and specialization improved on Zr-GO compared to Zr [49].

Conclusively, GO-coated zirconia hindered *S. mutans* bacterial attachment and promoted osteoblast growth and specialization, suggesting a potential prevention of peri-implantitis by deterring bacterial adhesion and enhancing implant success by improving bone attachment [49].

4.2. Scaffolds and Membranes

Recently, scientific research on graphene has focused on regenerative techniques in oral surgery, going on to investigate the efficacy of graphene oxide added to membranes or scaffolds compared with conventional methods with the hope that the results, combined with the potential of stem cells, will lead to a new class of nanomaterials with unique properties and a significant impact in the field of nanotechnology and oral health.

For example, the evaluation of a bone graft material consisting of biphasic calcium phosphate (BCP) coated with rGO was the focus of the investigation by Jeong-Woo Kim et al.

Osteoblast viability decreased as the concentration of rGO nanoplates increased in terms of cytotoxicity, with significant decreases at higher concentrations, while new bone production dramatically increased compared with the control group in in vivo tests using rat calvarial lesions. In fact, according to micro-CT and histomorphometric evaluations, rGO-coated BCP groups had higher volumes and percentages of new bone. The rGO4 group (with a 4:1000 ratio of rGO to BCP) showed the highest bone volume, demonstrating that the concentration of rGO in the composite material is important for bone regeneration [55].

In contrast, Erika Nishida et al. investigated the effects of adding a GO monolayer solution to a three-dimensional collagen scaffold for possible use in bone tissue engineering [56]. A special technique was used to produce the GO solution, and the resulting monolayer had an average width of about 20 m and a thickness of less than 1 nm. Next, the GO solution was mixed with a special solvent to obtain GO dispersions at various concentrations. The GO-modified scaffolds were injected into collagen scaffolds, and their different properties were evaluated. When the GO-modified scaffolds were characterized, they were found to have better physical characteristics, such as greater resistance to compression and enzymatic degradation, as well as a greater ability to adsorb calcium ions and proteins. To evaluate the effects of the modified GO scaffold, in vivo tests were performed on dog extractive cavities and rat subcutaneous tissues. The results showed that in rat tissues, the GO-modified scaffold stimulated angiogenesis and growth of cells and tissues and that compared with collagen-only scaffolds, the GO-modified scaffold significantly improved bone growth in dog extractive cavities. Overall, the research points to collagen scaffolds as attractive options for bone tissue engineering applications, as the addition of GO can improve their physical characteristics, cytocompatibility, and ability to form bone [56].

The research results of Jong Ho Lee et al. suggest that composite nanoparticles of hydroxyapatite (HAp) and rGO have extraordinary potential in enhancing the proliferation and osteogenic differentiation of pre-osteoblastic MC3T3-E1 cells.

In particular, extracellular calcium deposition in MC3T3-E1 cells was significantly enhanced by rGO/HAp composite nanoparticles, and clearly, calcium accumulation is a key sign of bone tissue creation and extracellular mineralization, two crucial steps of bone regeneration. This implies that composite nanoparticles could create a favorable environment for bone tissue development, thereby promoting osteogenesis. In addition, the enzymatic activity of alkaline phosphatase (ALP), an early indicator of osteogenic differentiation, was significantly elevated in the presence of composite nanoparticles. This shows that composite nanoparticles can accelerate the differentiation process of pre-osteoblastic cells.

The presence of composite nanoparticles also had a favorable impact on the deposition of osteogenic proteins, including osteopontin (OPN) and osteocalcin (OCN), which are essential markers of osteogenic cell development, suggesting a favorable impact on the expression of proteins important for bone growth. Therefore, these results also appear to be on the same wavelength as the other studies mentioned [57].

The work of Izumi Kanayama et al. [58] investigated the synthesis and characterization of GO and rGO films. Atomic force microscopy (AFM), scanning electron microscopy (SEM), and X-ray diffraction (XRD) were used to analyze the film morphology and tissue alterations of GO and rGO. Investigations on the biological characteristics of GO and rGO films were also conducted: the films were applied to culture plates and used by MC3T3-E1 mouse osteoblastic cells as substrates. From the results, it was found that the behavior of cells is affected differently by GO and rGO films. Compared with GO films, rGO films showed better cell activity. The films were also used to modify collagen scaffolds, resulting in improved tissue growth and compressive strength. Giant cells were present, and the materials and immune cells interacted favorably, indicating strong biocompatibility and a greater ability to stimulate cell activity and tissue integration. These results highlight the promising tissue engineering applications of GO and rGO and their ability to modify scaffolds to improve mechanical strength and tissue regeneration [58].

In the work of Chingis Daulbayev et al. [59], a composite was made by combining GO and HAp with a matrix of polycaprolactone (PCL), using an electrospinning technique, and the antibacterial capabilities of the composite on Gram-positive (*S. aureus*) and Gram-negative (*E. coli*) bacteria were analyzed; the antibacterial action of the composite was significant, and a larger clean zone was observed for higher concentrations of GO in the composite. The biocompatibility of the GO/HAp/PCL composite was also evaluated using MC3T3-E1 preosteoblast cells. Cell viability studies showed that the cytotoxic effects of the composite were minimal at lower concentrations: the cytotoxicity caused by HAp seemed to be attenuated by the addition of GO to the composite structure. It is also possible that GO has the ability to promote cell development, as it increases cell attachment and proliferation when added to the composite. Ultimately, the work showed that it was possible to successfully synthesize a GO/HAp/PCL composite with promising antibacterial qualities and biocompatibility [70–72].

In recently published work by Milena Radunovic et al. [60], the effects of GO-coated collagen membranes for guided bone regeneration (GBR) applications on dental pulp stem cells (DPSCs) were examined [73]. The research showed that attachment, proliferation, and osteogenic differentiation of DPSCs were promoted by the 2 and 10 g/mL GO-coated membranes, with a particular increase in metabolic activity, especially at higher concentrations. The fact that the cells adhered to the membrane surface without penetrating it was confirmed by hematoxylin-eosin staining. In addition, bone morphogenetic protein 2 (BMP2) and runt-related transcription factor 2 (RUNX2), indicators of osteogenic development, were significantly increased on GO-coated membranes, according to gene expression analysis. GO coating also significantly increased the secretion of prostaglandin E2 (PGE2), a crucial modulator of osteoblastic differentiation. On the other hand, with regard to inflammatory markers, tumor necrosis factor (TNF) and cyclo-oxygenase 2 (COX2) were downregulated, indicating a reduction in inflammation.

The study concluded that GO-coated collagen membranes limit inflammatory processes and promote attachment, proliferation, and osteogenic differentiation of DPSCs, emphasizing that efficacy is dose-dependent (the 10 g/mL concentration of GO produces the best results) [60].

The study presented by Letizia Ferroni et al. [61] investigated the creation and evaluation of rGO-PCL (reduced graphene oxide-polycaprolactone) composites for possible use in bone tissue engineering. Evaluation of the antibacterial efficacy of the composites against various bacterial strains revealed that they had a bacteriostatic effect on Gram-positive bacteria, particularly the 5% rGO-PCL composite. The study of the adhesion, morphology, and proliferation of human adipose-derived mesenchymal stem cells (HMSCs) was carried out on the surfaces of rGO-PCL, bringing out an upregulation of adhesion molecules, extracellular matrix elements (ECM) and metalloproteinases, indicating favorable cell-matrix interactions. In addition, the ability of HMSCs to differentiate into osteoblasts on the surfaces of rGO-PCL was examined. ALP activity and mineral matrix deposition were found to be maximal on the surface of 5% rGO-PCL, indicating better osteogenic differentiation capacity, and the expression of osteogenic markers such as OPN, OCN, RUNX2, and osterix (OSX) was found to be high on the surface of 5% rGO-PCL, indicating excellent osteoblastic proliferation. Certainly, therefore, the 5% rGO-PCL composite has proven to be a viable option for creating improved biomaterials for bone regeneration due to its demonstrated biocompatibility, bacteriostatic action against Gram-positive bacteria, and ability to enhance osteogenic differentiation [61].

In the very recent study by Elham-Sadat Motiee et al. [62] in 2023, a poly-3-hydroxybutyrate-chitosan (PC) scaffold reinforced with GO through the electrospinning method was developed with the aim of evaluating the fiber diameter, heat capacity, surface hydrophilicity, mechanical properties, and degradation of the nanocomposite scaffolds. It was again found that the above values are improved, suggesting that the improved characteristics and interactions of Poly-3 hydroxybutyrate-chitosan (PC)-GO nanocomposite scaffolds with cells and minerals may be promising for use in bone tissue engineering. In particular, the inclusion of GO improved the deposition of calcium and phosphate ions, indicating accelerated biomineralization, as well as increased cell adhesion, proliferation, and ALP activity, resulting in improved cell attachment, viability, and osteogenic activity [62].

The objective of the study by Alana P C Souza et al. [63] was to develop chitosan-xanthan-based membranes that also contained HAp and GO for potential use in guided bone regeneration (GBR) and again, as in the aforementioned studies, the results are extremely positive: in vitro bioactivity tests showed that HAp and GO increased bioactivity and promoted apatite deposition, and in particular higher concentrations of GO in membranes produced superior results in terms of cell viability, indicating increased cell adhesion and proliferation, which are essential for regenerative processes. Although the addition of particles did not improve mechanical properties, research on tensile strength showed that the membranes still exhibited qualities suitable for use as barriers and structural support in bone tissue regeneration [63].

4.3. Quality Assessment and Risk of Bias

The risk of bias in the included studies is reported in Figure 4. Regarding the randomization process, one study presents a high risk of bias and allocation concealment. All other studies ensure a low risk of bias. Only one study excludes performance; two studies confirm an increased risk of detection bias (self-reported outcome), and two of the included studies present a low detection bias (objective measures) (Figure 4). Two studies ensure a low risk regarding attrition and reporting bias.

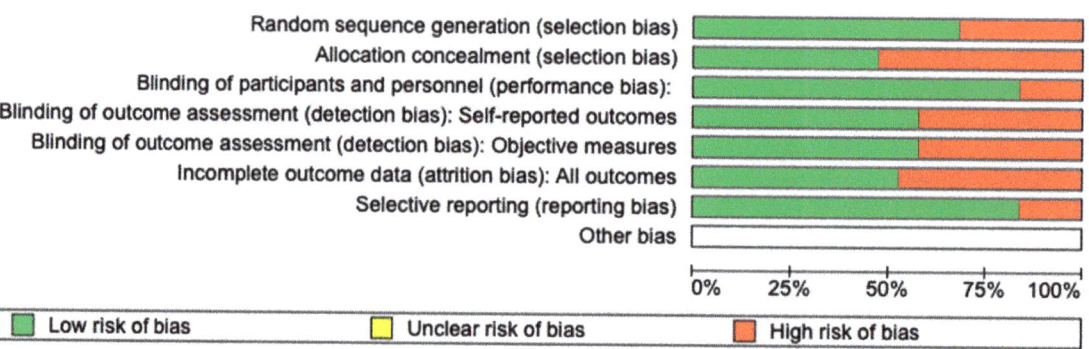

Figure 4. *Cont.*

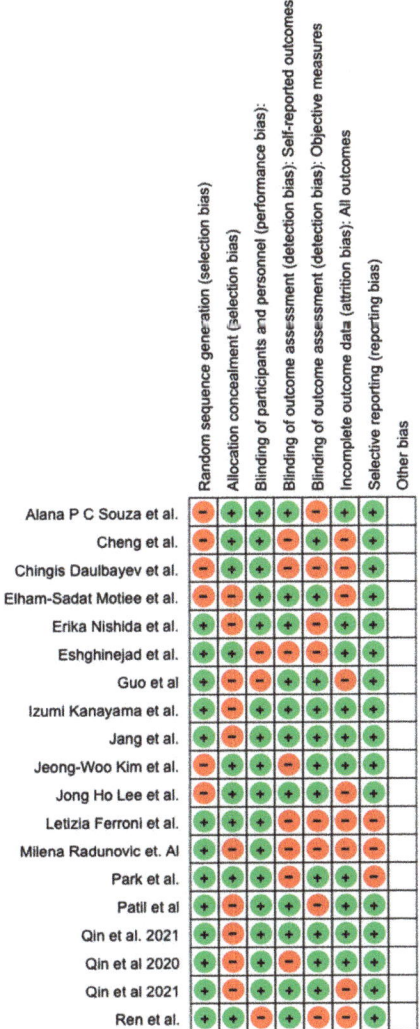

Figure 4. Risk of bias; red indicates high risk, and green indicates low risk of bias [45–63].

5. Conclusions

The field of graphene-based engineered nanomaterials in dentistry is expanding due to their superior mechanical properties, electrical conductivity, and thermal stability. Graphene and its derivatives can be functionalized with bioactive compounds and added to dental materials, enhancing their properties. These materials stimulate tissue regeneration, cell differentiation, and proliferation while being biocompatible and low in toxicity. Dental implants and abutments with graphene coatings exhibit improved cytocompatibility, antibacterial properties, and osteoblast growth.

Graphene-enhanced scaffolds and membranes for guided bone regeneration (GBR) also have improved physical properties, leading to enhanced bone formation. Laboratory tests indicate increased secretion of osteogenic markers and reduced inflammatory markers on graphene-coated materials. However, there are concerns about the safety of graphene and its derivatives, necessitating further research to understand their long-term effects.

Overall, graphene materials hold great potential for improving oral surgery procedures in the future.

Author Contributions: Conceptualization, A.M.I., A.P., G.L., G.P. and L.F.; methodology I.T., L.F., A.D.I., A.M.I., G.P. and A.P.; software, G.D., A.M.I., G.L. and A.P.; validation, I.T., F.I., G.L. and A.M.I.; formal analysis, A.M.I., A.D.I., F.I. and A.P.; investigation, G.D., L.F., A.M.I. and A.P.; resources, A.D.I., I.T., G.P, L.F., A.P. and F.L.; data curation, A.S., A.D.I., L.F., G.P., C.D.P. and A.P.; writing—original draft preparation, A.M.I., G.L., A.D.I., G.P. and A.P.; writing—review and editing, G.L., C.D.P., F.I. and A.P.; visualization, A.S., G.P., I.T., F.I. and G.L.; supervision, G.D., F.L., A.M.I., L.F., C.D.P. and F.I.; project administration, F.L., A.S., G.P., G.L., I.T. and A.P. All authors have read and agreed to the published version of the manuscript.

Funding: This research received no external funding.

Institutional Review Board Statement: Not applicable.

Informed Consent Statement: Not applicable.

Data Availability Statement: Not applicable.

Conflicts of Interest: The authors declare no conflict of interest.

Abbreviations

AFM	atomic force microscopy
ALP	alkaline phosphatase
AMP	antimicrobial peptide
BCP	biphasic calcium phosphate
BG	bioglass
BMP2	bone morphogenetic protein 2
CF	carbon fibers
COX2	cyclo-oxygenase 2
DEX	dexamethasone
DPSCs	dental pulp stem cells
ECM	extracellular matrix elements
FEA	finite element analysis
GBR	guided bone regeneration
GO	graphene oxide
HAp	hydroxyapatite
HGF	human gingival fibroblast
OCNr	osteocalcin
OPN	osteopontin
OSX	osterix
PC	poly-3-hydroxybutyrate-chitosan
PCL	polycaprolactone
PDA	poly-dopamine
PEEK	polyetheretherketone
PGE2	prostaglandin E2
PTT	photothermal therapy
rGO	reduced graphene oxide
rBMSCs	rat bone mesenchymal stem cells
RUNX2	runt-related transcription factor 2
SEM	scanning electron microscopy
Ti	titanium
TNF	tumor necrosis factor
XRD	X-ray diffraction
Zr	zirconia

References

1. Liao, C.; Xiao, S.; Wang, X. Bench-to-Bedside: Translational Development Landscape of Biotechnology in Healthcare. *Health Sci. Rev.* **2023**, *7*, 100097. [CrossRef]
2. Tiwari, S.K.; Sahoo, S.; Wang, N.; Huczko, A. Graphene Research and Their Outputs: Status and Prospect. *J. Sci. Adv. Mater. Devices* **2020**, *5*, 10–29. [CrossRef]
3. Yadav, S.; Raman, A.P.S.; Meena, H.; Goswami, A.G.; Bhawna; Kumar, V.; Jain, P.; Kumar, G.; Sagar, M.; Rana, D.K.; et al. An Update on Graphene Oxide: Applications and Toxicity. *ACS Omega* **2022**, *7*, 35387–35445. [CrossRef] [PubMed]
4. Sreenivasalu, P.K.P.; Dora, C.P.; Swami, R.; Jasthi, V.C.; Shiroorkar, P.N.; Nagaraja, S.; Asdaq, S.M.B.; Anwer, M.K. Nanomaterials in Dentistry: Current Applications and Future Scope. *Nanomaterials* **2022**, *12*, 1676. [CrossRef] [PubMed]
5. Yu, W.; Sisi, L.; Haiyan, Y.; Jie, L. Progress in the Functional Modification of Graphene/Graphene Oxide: A Review. *RSC Adv.* **2020**, *10*, 15328–15345. [CrossRef]
6. Allen, M.J.; Tung, V.C.; Kaner, R.B. Honeycomb Carbon: A Review of Graphene. *Chem. Rev.* **2010**, *110*, 132–145. [CrossRef]
7. Donato, K.Z.; Tan, H.L.; Marangoni, V.S.; Martins, M.V.S.; Ng, P.R.; Costa, M.C.F.; Jain, P.; Lee, S.J.; Koon, G.K.W.; Donato, R.K.; et al. Graphene Oxide Classification and Standardization. *Sci. Rep.* **2023**, *13*, 6064. [CrossRef] [PubMed]
8. Caffo, M.; Merlo, L.; Marino, D.; Caruso, G. Graphene in Neurosurgery: The Beginning of a New Era. *Nanomaterials* **2015**, *10*, 615–625. [CrossRef]
9. Tahriri, M.; Del Monico, M.; Moghanian, A.; Tavakkoli Yaraki, M.; Torres, R.; Yadegari, A.; Tayebi, L. Graphene and Its Derivatives: Opportunities and Challenges in Dentistry. *Mater. Sci. Eng. C Mater. Biol. Appl.* **2019**, *102*, 171–185. [CrossRef] [PubMed]
10. Srimaneepong, V.; Skallevold, H.E.; Khurshid, Z.; Zafar, M.S.; Rokaya, D.; Sapkota, J. Graphene for Antimicrobial and Coating Application. *Int. J. Mol. Sci.* **2022**, *23*, 499. [CrossRef] [PubMed]
11. Tabish, T.A.; Abbas, A.; Narayan, R.J. Graphene Nanocomposites for Transdermal Biosensing. *WIREs Nanomed. Nanobiotechnol.* **2021**, *13*, e1699. [CrossRef]
12. Htwe, Y.Z.N.; Mariatti, M. Printed Graphene and Hybrid Conductive Inks for Flexible, Stretchable, and Wearable Electronics: Progress, Opportunities, and Challenges. *J. Sci. Adv. Mater. Devices* **2022**, *7*, 100435. [CrossRef]
13. Mattei, T.A. How Graphene Is Expected to Impact Neurotherapeutics in the near Future. *Expert Rev. Neurother.* **2014**, *14*, 845–847. [CrossRef] [PubMed]
14. Calabriso, N.; Stanca, E.; Rochira, A.; Damiano, F.; Giannotti, L.; Di Chiara Stanca, B.; Massaro, M.; Scoditti, E.; Demitri, C.; Nitti, P.; et al. Angiogenic Properties of Concentrated Growth Factors (CGFs): The Role of Soluble Factors and Cellular Components. *Pharmaceutics* **2021**, *13*, 635. [CrossRef]
15. Li, X.; Liang, X.; Wang, Y.; Wang, D.; Teng, M.; Xu, H.; Zhao, B.; Han, L. Graphene-Based Nanomaterials for Dental Applications: Principles, Current Advances, and Future Outlook. *Front. Bioeng. Biotechnol.* **2022**, *10*, 804201. [CrossRef] [PubMed]
16. Ye, B.; Wu, B.; Su, Y.; Sun, T.; Guo, X. Recent Advances in the Application of Natural and Synthetic Polymer-Based Scaffolds in Musculoskeletal Regeneration. *Polymers* **2022**, *14*, 4566. [CrossRef]
17. Gianfreda, F.; Bollero, P. Dental Materials Design and Innovative Treatment Approach. *Dent. J.* **2023**, *11*, 85. [CrossRef]
18. Pavlíková, G.; Foltán, R.; Horká, M.; Hanzelka, T.; Borunská, H.; Sedý, J. Piezosurgery in Oral and Maxillofacial Surgery. *Int. J. Oral Maxillofac. Surg.* **2011**, *40*, 451–457. [CrossRef] [PubMed]
19. Chen, F.-M.; Liu, X. Advancing Biomaterials of Human Origin for Tissue Engineering. *Prog. Polym. Sci.* **2016**, *53*, 86–168. [CrossRef]
20. Lei, Y.; Zhang, T.; Lin, Y.C.; Granzier-Nakajima, T.; Bepete, G.; Kowalczyk, D.A.; Lin, Z.; Zhou, D.; Schranghamer, T.F.; Dodda, A.; et al. Graphene and beyond: Recent advances in two-dimensional materials synthesis, properties, and devices. *ACS Nanosci. Au* **2022**, *2*, 450–485. Available online: https://pubs.acs.org/doi/10.1021/acsnanoscienceau.2c00017 (accessed on 9 August 2023). [CrossRef]
21. Mantha, S.; Pillai, S.; Khayambashi, P.; Upadhyay, A.; Zhang, Y.; Tao, O.; Pham, H.M.; Tran, S.D. Smart Hydrogels in Tissue Engineering and Regenerative Medicine. *Materials* **2019**, *12*, 3323. [CrossRef] [PubMed]
22. Park, C.; Park, S.; Lee, D.; Choi, K.S.; Lim, H.-P.; Kim, J. Graphene as an Enabling Strategy for Dental Implant and Tissue Regeneration. *Tissue Eng. Regen. Med.* **2017**, *14*, 481–493. [CrossRef] [PubMed]
23. Lee, J.-H.; Park, S.-J.; Choi, J.-W. Electrical Property of Graphene and Its Application to Electrochemical Biosensing. *Nanomaterials* **2019**, *9*, 297. [CrossRef]
24. Sengupta, J.; Hussain, C.M. Graphene-Based Electrochemical Nano-Biosensors for Detection of SARS-CoV-2. *Inorganics* **2023**, *11*, 197. [CrossRef]
25. Costa, A.C.; Alves, P.M.; Monteiro, F.J.; Salgado, C. Interactions between Dental MSCs and Biomimetic Composite Scaffold during Bone Remodeling Followed by In Vivo Real-Time Bioimaging. *Int. J. Mol. Sci.* **2023**, *24*, 1827. [CrossRef] [PubMed]
26. Aryaei, A.; Jayatissa, A.H.; Jayasuriya, A.C. The Effect of Graphene Substrate on Osteoblast Cell Adhesion and Proliferation. *J. Biomed. Mater. Res. A* **2014**, *102*, 3282–3290. [CrossRef] [PubMed]
27. Qamar, Z.; Alghamdi, A.M.S.; Haydarah, N.K.B.; Balateef, A.A.; Alamoudi, A.A.; Abumismar, M.A.; Shivakumar, S.; Cicciù, M.; Minervini, G. Impact of Temporomandibular Disorders on Oral Health-related Quality of Life: A Systematic Review and Meta-analysis. *J. Oral Rehabil.* **2023**, *50*, 706–714. [CrossRef]

28. Mukherjee, S.P.; Bottini, M.; Fadeel, B. Graphene and the Immune System: A Romance of Many Dimensions. *Front. Immunol.* **2017**, *8*, 673. [CrossRef] [PubMed]
29. Inchingolo, A.D.; Malcangi, G.; Semjonova, A.; Inchingolo, A.M.; Patano, A.; Coloccia, G.; Ceci, S.; Marinelli, G.; Di Pede, C.; Ciocia, A.M.; et al. Oralbiotica/Oralbiotics: The Impact of Oral Microbiota on Dental Health and Demineralization: A Systematic Review of the Literature. *Children* **2022**, *9*, 1014. [CrossRef]
30. Ghenbot, Y.; Wathen, C.; Gutierrez, A.; Spadola, M.; Cucchiara, A.; Petrov, D. Effectiveness of Oral Antibiotic Therapy in Prevention of Postoperative Wound Infection Requiring Surgical Washout in Spine Surgery. *World Neurosurg.* **2022**, *163*, e275–e282. [CrossRef]
31. Hurdle, J.G.; O'Neill, A.J.; Chopra, I.; Lee, R.E. Targeting Bacterial Membrane Function: An Underexploited Mechanism for Treating Persistent Infections. *Nat. Rev. Microbiol.* **2011**, *9*, 62–75. [CrossRef]
32. Minetti, E.; Grassi, A.; Beca Campoy, T.; Palermo, A.; Mastrangelo, F. Innovative Alveolar Socket Preservation Procedure Using Demineralized Tooth Dentin as Graft Biomaterial Covered with Three Reabsorbable Membranes: Human Histological Case Series Evaluation. *Appl. Sci.* **2023**, *13*, 1411. [CrossRef]
33. Escobar, A.; Muzzio, N.; Moya, S.E. Antibacterial Layer-by-Layer Coatings for Medical Implants. *Pharmaceutics* **2021**, *13*, 16. [CrossRef] [PubMed]
34. Inchingolo, A.M.; Malcangi, G.; Inchingolo, A.D.; Mancini, A.; Palmieri, G.; Di Pede, C.; Piras, F.; Inchingolo, F.; Dipalma, G.; Patano, A. Potential of Graphene-Functionalized Titanium Surfaces for Dental Implantology: Systematic Review. *Coatings* **2023**, *13*, 725. [CrossRef]
35. Inchingolo, A.M.; Malcangi, G.; Ferrante, L.; Del Vecchio, G.; Viapiano, F.; Inchingolo, A.D.; Mancini, A.; Annicchiarico, C.; Inchingolo, F.; Dipalma, G.; et al. Surface Coatings of Dental Implants: A Review. *J. Funct. Biomater.* **2023**, *14*, 287. [CrossRef] [PubMed]
36. Greenhalgh, R.; Dempsey-Hibbert, N.C.; Whitehead, K.A. Antimicrobial Strategies to Reduce Polymer Biomaterial Infections and Their Economic Implications and Considerations. *Int. Biodeterior. Biodegrad.* **2019**, *136*, 1–14. [CrossRef]
37. Palermo, A.; Giannotti, L.; Stanca, B.D.C.; Ferrante, F.; Gnoni, A.; Nitti, P.; Calabriso, N.; Demitri, C.; Damiano, F.; Batani, T.; et al. Use of CGF in Oral and Implant Surgery: From Laboratory Evidence to Clinical Evaluation. *Int. J. Mol. Sci.* **2022**, *23*, 15164. [CrossRef]
38. Panich, M.; Poolthong, S. The Effect of Casein Phosphopeptide-Amorphous Calcium Phosphate and a Cola Soft Drink on In Vitro Enamel Hardness. *J. Am. Dent. Assoc.* **2009**, *140*, 455–460. [CrossRef]
39. Reina, G.; Iglesias, D.; Samorì, P.; Bianco, A. Graphene: A Disruptive Opportunity for COVID-19 and Future Pandemics? *Adv. Mater.* **2021**, *33*, e2007847. [CrossRef]
40. Balzanelli, M.G.; Distratis, P.; Dipalma, G.; Vimercati, L.; Catucci, O.; Amatulli, F.; Cefalo, A.; Lazzaro, R.; Palazzo, D.; Aityan, S.K.; et al. Immunity Profiling of COVID-19 Infection, Dynamic Variations of Lymphocyte Subsets, a Comparative Analysis on Four Different Groups. *Microorganisms* **2021**, *9*, 2036. [CrossRef]
41. Guazzo, R.; Gardin, C.; Bellin, G.; Sbricoli, L.; Ferroni, L.; Ludovichetti, F.S.; Piattelli, A.; Antoniac, I.; Bressan, E.; Zavan, B. Graphene-Based Nanomaterials for Tissue Engineering in the Dental Field. *Nanomaterials* **2018**, *8*, 349. [CrossRef] [PubMed]
42. Inchingolo, A.M.; Malcangi, G.; Ferrante, L.; Del Vecchio, G.; Viapiano, F.; Mancini, A.; Inchingolo, F.; Inchingolo, A.D.; Di Venere, D.; Dipalma, G.; et al. Damage from Carbonated Soft Drinks on Enamel: A Systematic Review. *Nutrients* **2023**, *15*, 1785. [CrossRef]
43. Carlson, N.E.; Roach, R.B. Platelet-Rich Plasma: Clinical Applications in Dentistry. *J. Am. Dent. Assoc.* **2002**, *133*, 1383–1386. [CrossRef]
44. Inchingolo, F.; Ballini, A.; Cagiano, R.; Inchingolo, A.; Serafini, M.; Benedittis, M.; Cortelazzi, R.; Tatullo, M.; Marrelli, M.; Inchingolo, A.M.; et al. Immediately Loaded Dental Implants Bioactivated with Platelet-Rich Plasma (PRP) Placed in Maxillary and Mandibular Region. *La Clin. Ter.* **2015**, *166*, e146–e152. [CrossRef]
45. Eshghinejad, P.; Farnoush, H.; Bahrami, M.S.; Bakhsheshi-Rad, H.R.; Karamian, E.; Chen, X.B. Electrophoretic Deposition of Bioglass/Graphene Oxide Composite on Ti-Alloy Implants for Improved Antibacterial and Cytocompatible Properties. *Mater. Technol.* **2020**, *35*, 69–74. [CrossRef]
46. Ren, N.; Li, J.; Qiu, J.; Yan, M.; Liu, H.; Ji, D.; Huang, J.; Yu, J.; Liu, H. Growth and Accelerated Differentiation of Mesenchymal Stem Cells on Graphene-Oxide-Coated Titanate with Dexamethasone on Surface of Titanium Implants. *Dent. Mater.* **2017**, *33*, 525–535. [CrossRef] [PubMed]
47. Park, L.; Kim, H.-S.; Jang, W.; Ji, M.-K.; Ryu, J.-H.; Cho, H.; Lim, H.-P. Antibacterial Evaluation of Zirconia Coated with Plasma-Based Graphene Oxide with Photothermal Properties. *Int. J. Mol. Sci.* **2023**, *24*, 8888. [CrossRef] [PubMed]
48. Cheng, Q.; Lu, R.; Wang, X.; Chen, S. Antibacterial Activity and Cytocompatibility Evaluation of the Antimicrobial Peptide Nal-P-113-Loaded Graphene Oxide Coating on Titanium. *Dent. Mater. J.* **2022**, *41*, 905–915. [CrossRef] [PubMed]
49. Jang, W.; Kim, H.-S.; Alam, K.; Ji, M.-K.; Cho, H.-S.; Lim, H.-P. Direct-Deposited Graphene Oxide on Dental Implants for Antimicrobial Activities and Osteogenesis. *Int. J. Nanomed.* **2021**, *16*, 5745–5754. [CrossRef] [PubMed]
50. Guo, C.; Lu, R.; Wang, X.; Chen, S. Graphene Oxide-Modified Polyetheretherketone with Excellent Antibacterial Properties and Biocompatibility for Implant Abutment. *Macromol. Res.* **2021**, *29*, 351–359. [CrossRef]

51. Qin, W.; Ma, J.; Liang, Q.; Li, J.; Tang, B. Tribological, Cytotoxicity and Antibacterial Properties of Graphene Oxide/Carbon Fibers/Polyetheretherketone Composite Coatings on Ti-6Al-4V Alloy as Orthopedic/Dental Implants. *J. Mech. Behav. Biomed. Mater.* **2021**, *122*, 104659. [CrossRef] [PubMed]
52. Qin, W.; Li, Y.; Ma, J.; Liang, Q.; Cui, X.; Jia, H.; Tang, B. Osseointegration and Biosafety of Graphene Oxide Wrapped Porous CF/PEEK Composites as Implantable Materials: The Role of Surface Structure and Chemistry. *Dent. Mater.* **2020**, *36*, 1289–1302. [CrossRef] [PubMed]
53. Qin, W.; Wang, C.; Jiang, C.; Sun, J.; Yu, C.; Jiao, T. Graphene Oxide Enables the Reosteogenesis of Previously Contaminated Titanium In Vitro. *J. Dent. Res.* **2020**, *99*, 922–929. [CrossRef] [PubMed]
54. Patil, V.; Naik, N.; Gadicherla, S.; Smriti, K.; Raju, A.; Rathee, U. Biomechanical Behavior of Bioactive Material in Dental Implant: A Three-Dimensional Finite Element Analysis. *Sci. World J.* **2020**, *2020*, e2363298. [CrossRef]
55. Kim, J.-W.; Shin, Y.C.; Lee, J.-J.; Bae, E.-B.; Jeon, Y.-C.; Jeong, C.-M.; Yun, M.-J.; Lee, S.-H.; Han, D.-W.; Huh, J.-B. The Effect of Reduced Graphene Oxide-Coated Biphasic Calcium Phosphate Bone Graft Material on Osteogenesis. *Int. J. Mol. Sci.* **2017**, *18*, 1725. [CrossRef]
56. Nishida, E.; Miyaji, H.; Kato, A.; Takita, H.; Iwanaga, T.; Momose, T.; Ogawa, K.; Murakami, S.; Sugaya, T.; Kawanami, M. Graphene Oxide Scaffold Accelerates Cellular Proliferative Response and Alveolar Bone Healing of Tooth Extraction Socket. *Int. J. Nanomed.* **2016**, *11*, 2265–2277. [CrossRef]
57. Lee, J.H.; Shin, Y.C.; Lee, S.-M.; Jin, O.S.; Kang, S.H.; Hong, S.W.; Jeong, C.-M.; Huh, J.B.; Han, D.-W. Enhanced Osteogenesis by Reduced Graphene Oxide/Hydroxyapatite Nanocomposites. *Sci. Rep.* **2015**, *5*, 18833. [CrossRef]
58. Kanayama, I.; Miyaji, H.; Takita, H.; Nishida, E.; Tsuji, M.; Fugetsu, B.; Sun, L.; Inoue, K.; Ibara, A.; Akasaka, T.; et al. Comparative Study of Bioactivity of Collagen Scaffolds Coated with Graphene Oxide and Reduced Graphene Oxide. *Int. J. Nanomed.* **2014**, *9*, 3363–3373. [CrossRef]
59. Daulbayev, C.; Sultanov, F.; Korobeinyk, A.V.; Yeleuov, M.; Taurbekov, A.; Bakbolat, B.; Umirzakov, A.; Baimenov, A.; Daulbayev, O. Effect of Graphene Oxide/Hydroxyapatite Nanocomposite on Osteogenic Differentiation and Antimicrobial Activity. *Surf. Interfaces* **2022**, *28*, 101683. [CrossRef]
60. Radunovic, M.; De Colli, M.; De Marco, P.; Di Nisio, C.; Fontana, A.; Piattelli, A.; Cataldi, A.; Zara, S. Graphene Oxide Enrichment of Collagen Membranes Improves DPSCs Differentiation and Controls Inflammation Occurrence. *J. Biomed. Mater. Res. Part A* **2017**, *105*, 2312–2320. [CrossRef]
61. Ferroni, L.; Gardin, C.; Rigoni, F.; Balliana, E.; Zanotti, F.; Scatto, M.; Riello, P.; Zavan, B. The Impact of Graphene Oxide on Polycaprolactone PCL Surfaces: Antimicrobial Activity and Osteogenic Differentiation of Mesenchymal Stem Cell. *Coatings* **2022**, *12*, 799. [CrossRef]
62. Motiee, E.-S.; Karbasi, S.; Bidram, E.; Sheikholeslam, M. Investigation of Physical, Mechanical and Biological Properties of Polyhydroxybutyrate-Chitosan/Graphene Oxide Nanocomposite Scaffolds for Bone Tissue Engineering Applications. *Int. J. Biol. Macromol.* **2023**, *247*, 125593. [CrossRef] [PubMed]
63. Souza, A.P.C.; Neves, J.G.; Navarro Da Rocha, D.; Lopes, C.C.; Moraes, Â.M.; Correr-Sobrinho, L.; Correr, A.B. Chitosan/Xanthan Membrane Containing Hydroxyapatite/Graphene Oxide Nanocomposite for Guided Bone Regeneration. *J. Mech. Behav. Biomed. Mater.* **2022**, *136*, 105464. [CrossRef] [PubMed]
64. Maiorana, C. Histomorphometric Evaluation of Anorganic Bovine Bone Coverage to Reduce Autogenous Grafts Resorption: Preliminary Results. *Open Dent. J.* **2011**, *5*, 71–78. [CrossRef]
65. Lorusso, F.; Inchingolo, F.; Greco Lucchina, A.; Scogna, G.; Scarano, A. Graphene-Doped Poly(Methyl-Methacrylate) as an Enhanced Biopolymer for Medical Device and Dental Implant. *J. Biol. Regul. Homeost. Agents* **2021**, *35*, 195–204. [CrossRef]
66. Liu, C.; Tan, D.; Chen, X.; Liao, J.; Wu, L. Research on Graphene and Its Derivatives in Oral Disease Treatment. *Int. J. Mol. Sci.* **2022**, *23*, 4737. [CrossRef] [PubMed]
67. Ghosal, K.; Sarkar, K. Biomedical Applications of Graphene Nanomaterials and Beyond. *ACS Biomater. Sci. Eng.* **2018**, *4*, 2653–2703. [CrossRef]
68. Ferrari, A.C.; Bonaccorso, F.; Fal'ko, V.; Novoselov, K.S.; Roche, S.; Bøggild, P.; Borini, S.; Koppens, F.H.L.; Palermo, V.; Pugno, N.; et al. Science and Technology Roadmap for Graphene, Related Two-Dimensional Crystals, and Hybrid Systems. *Nanoscale* **2015**, *7*, 4598–4810. [CrossRef]
69. Sharifian Jazi, F.; Khaksar, S.; Esmaeilkhanian, A.; Bazli, L.; Eskandarinezhad, S.; Salahshour, P.; Sadeghi, F.; Rostamnia, S. Advancements in Fabrication and Application of Chitosan Composites in Implants and Dentistry: A Review. *Biomolecules* **2022**, *12*, 155. [CrossRef]
70. Minervini, G.; Franco, R.; Marrapodi, M.M.; Di Blasio, M.; Isola, G.; Cicciù, M. Conservative Treatment of Temporomandibular Joint Condylar Fractures: A Systematic Review Conducted According to PRISMA Guidelines and the Cochrane Handbook for Systematic Reviews of Interventions. *J. Oral Rehabil.* **2023**, *50*, 886–893. [CrossRef]
71. Minervini, G.; Franco, R.; Marrapodi, M.M.; Crimi, S.; Badnjević, A.; Cervino, G.; Bianchi, A.; Cicciù, M. Correlation between Temporomandibular Disorders (TMD) and Posture Evaluated Trough the Diagnostic Criteria for Temporomandibular Disorders (DC/TMD): A Systematic Review with Meta-Analysis. *J. Clin. Med.* **2023**, *12*, 2652. [CrossRef] [PubMed]

72. Minervini, G.; Franco, R.; Marrapodi, M.M.; Fiorillo, L.; Cervino, G.; Cicciù, M. Economic Inequalities and Temporomandibular Disorders: A Systematic Review with Meta-analysis. *J. Oral Rehabil.* **2023**, *50*, 715–723. [CrossRef] [PubMed]
73. Inchingolo, F.; Tatullo, M.; Abenavoli, F.M.; Marrelli, M.; Inchingolo, A.D.; Gentile, M.; Inchingolo, A.M.; Dipalma, G. Non-Syndromic Multiple Supernumerary Teeth in a Family Unit with a Normal Karyotype: Case Report. *Int. J. Med. Sci.* **2010**, *7*, 378–384. [CrossRef] [PubMed]

Disclaimer/Publisher's Note: The statements, opinions and data contained in all publications are solely those of the individual author(s) and contributor(s) and not of MDPI and/or the editor(s). MDPI and/or the editor(s) disclaim responsibility for any injury to people or property resulting from any ideas, methods, instructions or products referred to in the content.

Article

Gold Nanoparticles AuNP Decorated on Fused Graphene-like Materials for Application in a Hydrogen Generation

Erik Biehler, Qui Quach and Tarek M. Abdel-Fattah *

Applied Research Center, Thomas Jefferson National Accelerator Facility, Department of Molecular Biology and Chemistry, Christopher Newport University, Newport News, VA 23606, USA; erik.biehler.16@cnu.edu (E.B.); qui.quach.13@cnu.edu (Q.Q.)
* Correspondence: fattah@cnu.edu

Abstract: The search for a sustainable, alternative fuel source to replace fossil fuels has led to an increased interest in hydrogen fuel. This combustible gas is not only clean-burning but can readily be produced via the hydrolysis of sodium borohydride. The main drawback of this reaction is that the reaction occurs relatively slowly and requires a catalyst to improve efficiency. This study explored a novel composite material made by combining gold nanoparticles and fused graphene-like materials (AuFGLM) as a catalyst for generating hydrogen via sodium borohydride. The novel fused graphene-like material (FGLM) was made with a sustainable dextrose solution and by using a pressure-processing method. Imaging techniques showed that FGLM appears to be an effective support template for nanoparticles. Transmission electron microscopy (TEM), scanning electron microscopy (SEM), energy-dispersive X-ray spectroscopy (EDS), Fourier-transform infrared spectroscopy (FTIR), X-ray diffraction (XRD) and Raman spectroscopy were used to characterize and determine the size, shape, and structure of nanoparticles and composites. The TEM study characterized the fused carbon backbone as it began to take on a rounder shape. The TEM images also revealed that the average diameter of the gold nanoparticle was roughly 23 nm. The FTIR study confirmed O-H, C-C, and C=O as functional groups in the materials. The EDS analysis showed that the composite contained approximately 6.3% gold by weight. The crystal structures of FGLM and AuFGLM were identified via P-XRD analysis. Various reaction conditions were used to test the catalytic ability of AuFGLM, including various solution pHs, temperatures, and doses of $NaBH_4$. It was observed that optimal reaction conditions included high temperature, an acidic solution pH, and a higher dose of $NaBH_4$. The activation energy of the reaction was determined to be 45.5 kJ mol^{-1}, and it was found that the catalyst could be used multiple times in a row with an increased volume of hydrogen produced in ensuing trials. The activation energy of this novel catalyst is competitive compared to similar catalysts and its ability to produce hydrogen over multiple uses makes the material an exciting choice for catalyzing the hydrolysis of $NaBH_4$ for use as a hydrogen fuel source.

Keywords: gold nanoparticles; fused graphene-like materials; composites; catalyst; hydrogen generation

1. Introduction

It is estimated that up to 84.3% of global energy comes from fossil fuels. These fossil fuels are not only dwindling, but their combustion is a leading cause of greenhouse gas emissions [1,2]. These issues have led to a search for an alternative energy source to replace the world's dependence on fossil fuels. One promising option is that of hydrogen gas, which is not only the most abundant element in the universe but is also clean-burning, as it only produces water as a byproduct. Safety concerns regarding the storage of this highly combustible gas are standing in the way of the widespread implementation of hydrogen as a fuel source; thus, much work is currently being done to find a safer method of storage. One potential method is the use of a hydrogen feedstock material (HFM), particularly sodium borohydride (NaBH$_4$), which contains 10.8% hydrogen by weight. NaBH$_4$ also readily reacts in water, releasing gaseous hydrogen (1) [3].

$$NaBH_4 + 2H_2O \rightarrow NaBO_2 + 4H_2 \qquad (1)$$

This reaction, however, occurs relatively slowly and requires a catalyst to improve the generation rate of hydrogen [3]. In recent years, precious metal nanoparticles have been utilized for a variety of different applications including gas detection [4], anti-bacterial properties [5], optoelectrical properties [6], as catalysts [7,8], and for hydrogen generation [9,10], including as a catalyst for the hydrolysis of NaBH$_4$ [11–14]. Gold nanoparticles, for instance, are used as catalysts in various chemical reactions. Although gold nanoparticles are not as widely used as other transition metal nanoparticles, such as palladium or platinum, gold nanoparticles possess unique properties that make them valuable catalysts in specific applications. Gold nanoparticles exhibit excellent catalytic activity and selectivity, particularly in oxidation and reduction reactions [15–19]. They can effectively catalyze reactions involving small molecules, such as carbon monoxide, oxygen, hydrogen, and alcohols. The catalytic properties of gold nanoparticles are highly dependent on their size and shape [20]. By controlling these parameters during synthesis, researchers can tailor the catalytic performance of the nanoparticles for specific reactions. Gold nanoparticles can be synthesized using several methods, including chemical reduction, sol–gel processes, and colloidal synthesis [15–20]. Smaller-sized nanoparticles often exhibit higher catalytic activity, especially when catalyzing the hydrolysis of NaBH$_4$. For example, in a study by Quach et al. (2021), the researchers synthesized gold nanoparticles with an average size of 8 nm and applied them as catalysts in a hydrogen generation reaction [11]. The catalyzing reactions achieved the highest reaction rate of pH 6, 303 K with 1225 μmoles of NaBH$_4$ [11]. The gold nanoparticles successfully kept catalyzing the reaction up to the fifth trial and assisted the reaction in generating 28.9 mL of hydrogen [12].

It is important to note that one drawback of nanoparticles is their tendency to agglomerate in solutions, which can lead to a decrease in their total surface area and overall catalytic ability [21]. Therefore, these nanoparticle catalysts are often combined with a support structure. The support structure not only increases the stability of the nanoparticles but also lowers the required activation energy. For example, a study by Osborne et al. (2020) indicated that the activation energy of gold nanoparticles supporting over-activated carbon (21.6 kJ mol^{-1}) was much lower than the unsupported nanoparticles (231.7 kJ mol^{-1}) [22]. It is challenging to find a safe, effective, and economical supported material; however, various support templates have been explored, including carbon nanotubes (CNTs). Graphene and mesoporous silica have been used to stabilize and separate nanoparticles [23]. Carbon-based materials are often an ideal choice for support materials due to their unique properties, such as a high surface area and stability [21–25]. One example of a carbon-based material that could potentially be used as a support for nanoparticles is, graphene, a well-known material which has been applied in the fields of batteries, solar cells, sensors, and catalysis [23]. Graphene is among the thinnest materials, and possesses exceptional mechanical strength. Its structure contains two-dimensional sp^2-hybridized carbon atom planar sheets stacked into honeycomb lattices. It was reported in study by Ghosh et al. that the surface area of graphene was higher than that of CNTs [26]. However, the production cost of pristine graphene is high, and its synthesis method is quite difficult [23]. Other graphene structures, including graphene oxide and reduced graphene oxide, have a lower production cost and easier preparation methods, but their methods of production often required toxic chemicals that pose a risk to human health [23,27]. In our study, we found a novel way to synthesize a fused graphene-like material (FGLM). The method was simple and only used dextrose, a renewable and eco-friendly material. The synthesized FGLM was then used as a support material in order to prevent gold nanoparticle agglomeration and improve their catalytic ability.

In this study, the reduction method was used to control the size, shape, and surface properties of the gold nanoparticles, positively influencing their catalytic behavior. Next, the fused graphene-like material was synthesized, characterized, and used as a support structure. The two synthesized materials were then combined to form the novel composite material, which was comprised of gold nanoparticles supported on fused graphene-like materials (AuFGLM). This new material was characterized and explored for its ability to catalyze the hydrolysis reaction of $NaBH_4$. The AuFGLM material was characterized via transmission electron microscopy (TEM), energy dispersive spectroscopy (EDS), Fourier transform infrared spectroscopy (FTIR), X-ray diffraction (XRD), and Raman spectroscopy. Then, the AuFGLM was tested for its catalytic ability under different doses of $NaBH_4$, at different temperatures, and pH values, and to determine its reusability.

2. Experimental
2.1. Synthesis of Gold Nanoparticles

We applied the method of synthesizing nanoparticles by described by Quach and Abdel-Fattah [5]. Gold nanoparticles were independently formed from the reduction of chloroauric acid ($AuCl_4H$) using sodium citrate as a stabilizer. First a 1 mM precursor solution was prepared by dissolving the chloroauric acid in 100 mL of water. This solution was then heated until boiling and 1% w/w sodium citrate solution was added dropwise until a color change was observed. The solution was removed from the heat and allowed to cool to room temperature.

2.2. Synthesis of Fused Graphene-like Materials and Nanocomposites

To synthesize the fused graphene-like material, a 0.5 M dextrose solution was prepared by dissolving powdered dextrose into deionized water (18 MΩ). The dextrose solution was added to a stainless-steel reaction vessel in a ratio of 3:2, gas to liquid. The reaction vessel was then placed into an oven and heated at 200 °C for four hours to form the fused graphene-like materials (FGLM). The FGLM material was centrifuged and washed several times with deionized water (18 MΩ) and ethanol to remove any impurities. Once washed, the fused graphene-like materials were dried and stored for future use.

The AuFGLM composites were produced by incipient wetness impregnation of roughly 300 mg FGLMs with 50 mL of the previously prepared gold nanoparticle solution containing 9.3 mg AuNPs. The two components were then mixed together at room temperature, and the produced material was then stored in an oven at 60 °C until all the liquid evaporated, as shown in Scheme 1.

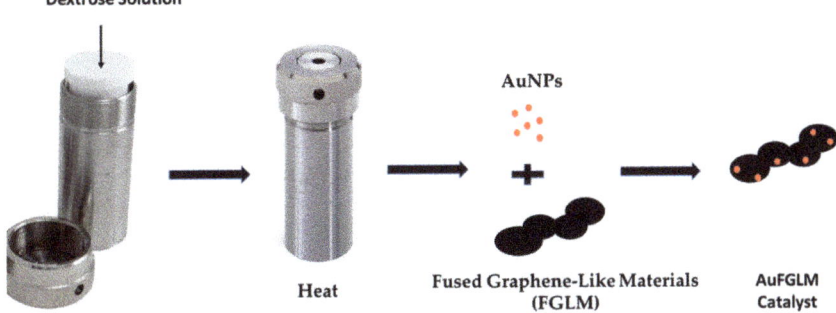

Scheme 1. Illustration of AuFGLM composite synthesis. The black arrow depicted the step by step of the synthesis process andhe red dots represent gold nanoparticles (AuNPs).

The resulting composite material was then characterized using transmission electron microscopy (TEM), scanning electron microscopy (SEM), energy-dispersive X-ray spectroscopy (EDS), Fourier-transform infrared spectroscopy (FTIR) and X-ray diffraction (XRD).

2.3. Characterization

For the transmission electron microscopy analysis, a sample of 10 mg of the AuFGLM catalyst was dispersed and sonicated in 5 mL of DI water. Therefore, the final concentration of the composite dispersion prepared for the TEM grid was 2 g per liter. A few drops of the resulting solution were added to a copper grid (300 mesh). Several of these grids were made and dried in an oven at 80 °C overnight. Each grid was then staged and scanned via transmission electron microscopy (TEM, JEM-2100F).

Images of the AuFGLM material were also obtained by scanning electron microscopy (SEM, JEOL JSM-6060LV, Akishima, Tokyo, Japan). This analysis was coupled with energy-dispersive X-ray spectroscopy (EDS, Thermo Scientific UltraDry, Waltham, MA, USA), allowing us to determine the weight ratio of the different elements present within the composite.

The composite's crystal structures were characterized using powdered X-ray diffraction (P-XRD, Rigaku Miniflex II, Tokyo, Japan), Cu Kα X-ray, nickel filters). Each sample was spread flat on a sample holder and inserted into the instrument which scanned the material from 5° to 90°.

Fourier-transform infrared spectroscopy (FTIR, Shimadzu IR-Tracer 100, Kyoto, Japan) with an attenuated total reflectance attachment (ATR, Shimadzu QATR-S, Kyoto, Japan) was then used to determine any functional groups present in the sample with a scanning range from 4000 cm^{-1} to 500 cm^{-1}.

The chemical structure of the AuGLM was also verified by Raman microscope and spectrometer (Renishaw, ISC3-1233, Wotton-under-Edge, UK). The sample was spread flat on the sample slide. The AuFGLM was first scanned at 50× focus from 2000 cm^{-1} to 150 cm^{-1}. Then, the microscope was changed to 100× focus to scan the AuNPs on the AuFGLM from 2000 cm^{-1} to 150 cm^{-1}.

2.4. Catalysis

The setup consisted of two vacuum flasks connected by a simple plastic hose. One flask was designated as the reaction chamber inside of which the hydrolysis reaction of $NaBH_4$ catalyzed by the novel AuFGLM catalyst occurred. The second flask contained DI water to be displaced by the hydrogen generated in the first flask. Both reaction chambers were sealed using rubber stoppers; however, the rubber stopper used to seal the second flask, containing the DI water to be displaced, was connected to a cup on a mass balance via a hose that went through the rubber stopper. This mass balance was connected to a computer which recorded the measured mass of the water displaced by the hydrogen gas every 0.25 s. This experiment was run at a variety of pH solutions (6, 7, 8), temperatures (283 K, 288 K, 295 K, 303 K), and $NaBH_4$ doses (625 µmoles, 925 µmoles, 1225 µmoles) in order to determine which reaction conditions were optimal. All reactions were stirred using a magnetic stir bar for a full two hours of the trial.

3. Results and Discussion

TEM micrographs (Figure 1) were used to confirm the presence of nanoparticles supported on t fused graphene-like materials. The image clearly shows the fused carbon backbone as it begins to take on a rounder shape. The material appears to have multiple thin layers stacked upon each other with areas of overlap that appear darker. Dark spherical gold nanoparticles can be seen on the material; the average diameter of the nanoparticles was determined to be 15 nm (±0.5 nm).

The AuFGLM catalyst was then characterized using EDS analysis. The EDS analysis (Figure 2) indicated that the AuFGLM contained approximately 49% carbon (wt%), 49% oxygen (wt%), and 2% gold (wt%).

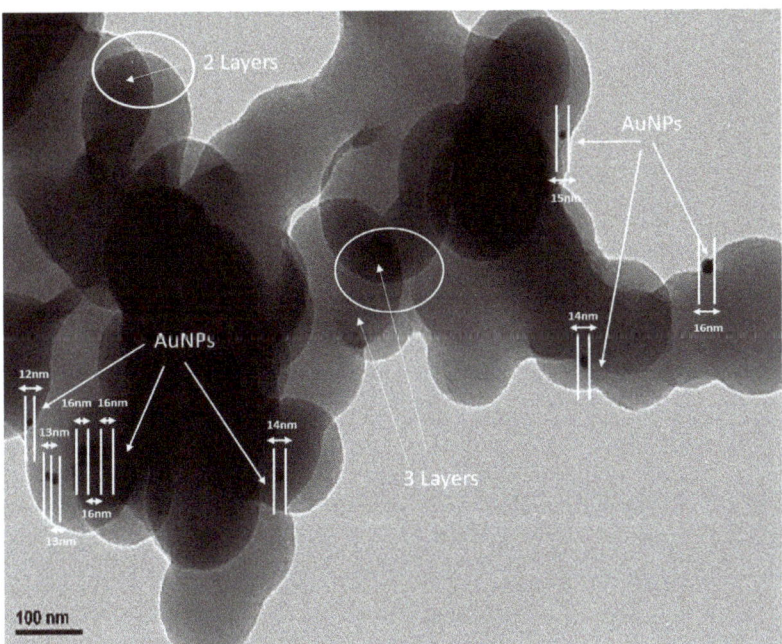

Figure 1. Transmission electron microscopy image of the novel AuFGLM composite catalyst at a scanning range of 100 nm. The white circle depicted the layers of AuFGLM.

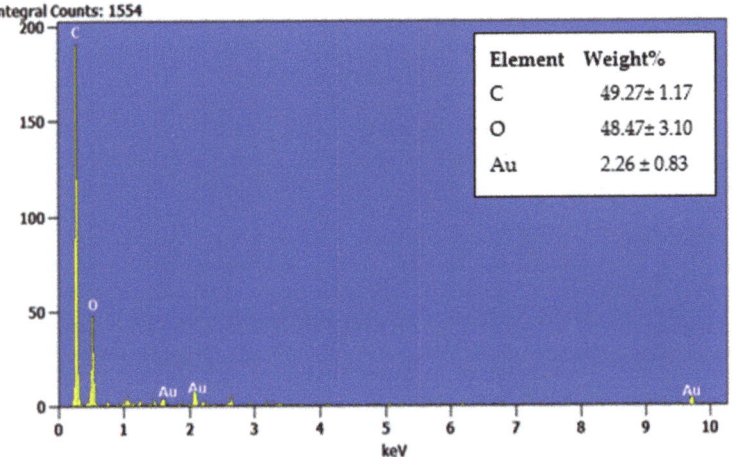

Figure 2. Energy-dispersive X-ray spectroscopy analysis for the novel AuFGLM catalyst.

The low weight percentage of the gold indicated that either the size of the nanoparticles was incredibly small or the gold was very well-dispersed throughout the sample. Due to the low weight percentage, the XRD peaks of the gold nanoparticles in Figure 3 were smaller than that of carbon. The small size of the nanoparticles is supported by the TEM images shown in Figure 1.

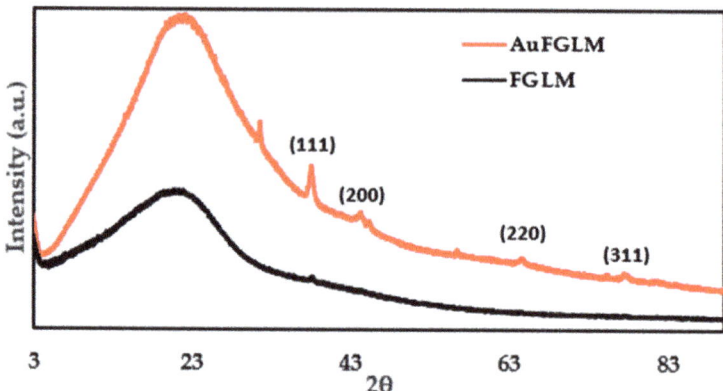

Figure 3. Powder X-ray diffraction patterns of the unsupported fused graphene-like materials and the novel AuFGLM composite.

The FGLM and AuFGLM materials were also characterized using powder X-ray diffraction (Figure 3). In both materials, a large broad peak can be observed at 23 degrees. This peak is commonly associated with carbon-based materials and is indicative of the fused graphene-like materials [24]. Four peaks that are indicative of gold nanoparticles were seen at 38.2°, 44.5°, 65.3°, and 77.7°. These peaks can, respectively, be attributed to the (111), (200), (220) and (311) lattice planes of the gold nanoparticles present in the material (JCPDS 04-0784) [28].

Figure 4 showed the Raman spectra of AuFGLM at 50× and 100× focus. At 50× focus, the AuFGLM indicated the G band at 1600 cm^{-1} and D band at 1374 cm^{-1} of the FGLM. A study by Perumbilavil et al. (2015) reported that the G band of graphene materials ranged from 1607 cm^{-1} to 1595 cm^{-1} [29]. The G band occurs as a result of the sp^2 carbon form. In a study by Lee et al. (2021), the graphene materials showed that the D band ranged from 1330 cm^{-1} to 1380 cm^{-1} [30]. The D band is formed from the defects and disorder in the carbon lattice. Since the AuNPs were well-dispersed on the materials, we turned the microscope to 100× to focus the laser on the AuNPs. The inset in Figure 4 shows that the AuNPs have Raman shifts at 1095 cm^{-1}, 784 cm^{-1} and 554 cm^{-1}. The results are approximately consistent with many reported studies. For example, a study by Govindaraju et al. (2015) reported that the Raman shifts of gold nanoparticles were found to be between 419 cm^{-1} and 709 cm^{-1} [31]. In a study by Lai et al. (2017), Raman shifts between 1000–1100 cm^{-1} were observed in gold nanoparticles [32].

The fused graphene-like material and AuFGLM composite materials were characterized using FTIR (Figure 5). The fused graphene-like material exhibits a long broad peak from roughly 3600 cm^{-1} to 3000 cm^{-1} which is characteristic of a hydroxyl functional group. A simple alkane group is represented by the short peak found at 2900 cm^{-1}. A peak seen at 1658 cm^{-1} is indicative of a C=O functional group. All three of these functional groups are characteristic of the dextrose molecule used to synthesize the fused graphene-like material backbone. As expected, all three of these peaks were also seen in the AuFGLM composite with little to no change in intensity, and with small shifts due to the presence of AnNPs. This indicates that the addition of gold nanoparticles to the fused graphene-like material backbone did not significantly alter its structure.

The catalytic efficiency of AuFGLM began by testing its performance at different doses of NaBH$_4$ (Figure 6). The catalyst was first tested at a dose of 625 μmol of NaBH$_4$, which produced roughly 15.8 mL of hydrogen after a trial time of two hours. The hydrogen generation rate was then calculated to be 0.0132 mL min^{-1} mg$_{cat}^{-1}$. Increasing the dose of sodium borohydride to 925 μmol resulted in an increased hydrogen generation rate of 0.0180 mL min^{-1} mg$_{cat}^{-1}$ and a volume of 21.6 mL of hydrogen gas produced. Further increasing the dose to 1225 µmol increased the rate and volume of hydrogen generated even further, at 0.0234 mL min^{-1} mg$_{cat}^{-1}$ and 28.1 mL, respectively. The data show that there is a direct relationship between the dosage of NaBH$_4$ used and the volume of hydrogen generated, which agrees with Le Chatlier's principle and Equation (3).

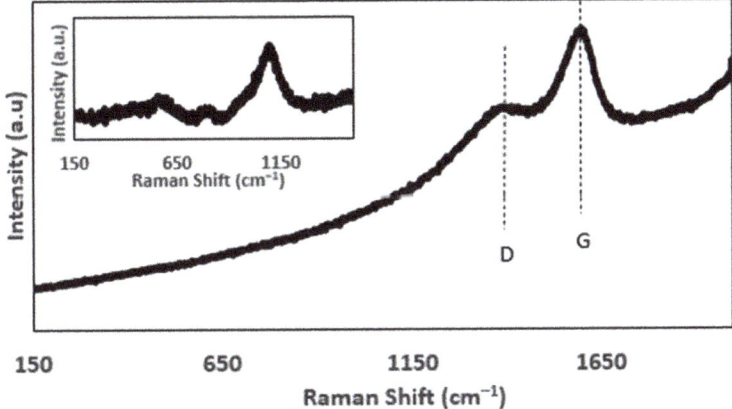

Figure 4. Raman spectrum of AuFGLM at 50× focus. The inset shows the Raman spectrum of AuFGLM at 100× focus. D indicated the D band. G indicated the G band.

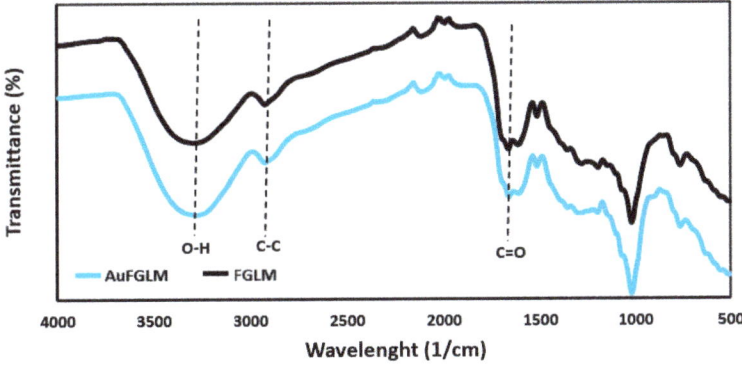

Figure 5. Fourier-transform infrared spectrum of the FGLM support material and the novel AuFGLM composite.

Next, the catalytic ability of AuFGLM was tested under various pH solutions, as shown in Figure 7. First, the reaction was tested at neutral conditions (pH 7), which produced a hydrogen gas volume of 21.6 mL at a rate of 0.0180 mL min^{-1}mg$_{cat}^{-1}$. The pH was then lowered to pH 6 which resulted in over twice as much hydrogen gas, with 53.5 mL produced at a rate of 0.0446 mL min^{-1}mg$_{cat}^{-1}$. When the pH of the reaction was raised to pH 8, it was observed that the rate of hydrogen generation had decreased to 0.0118 mL min^{-1}mg$_{cat}^{-1}$, producing only 14.1 mL of hydrogen gas after two hours. Based on this data, it was determined that the reaction catalyzed by AuFGLM performed better under acidic conditions. This phenomenon was previously reported in work by Schlesinger et al. (1953) who found that stronger acids sped up the hydrogen generation of this reaction, possibly due to an increase in the free hydrogen ions present in the solution [3]. Inversely, increasing the pH of the reaction was observed to slow the rate of generation, as was observed by Kaufman et. al. (1985) and also by Grzeschik et al. (2020), and in previous work by this research team [11–14,33–37]. It may be that hydroxide ions compete for hydrogen in the solution.

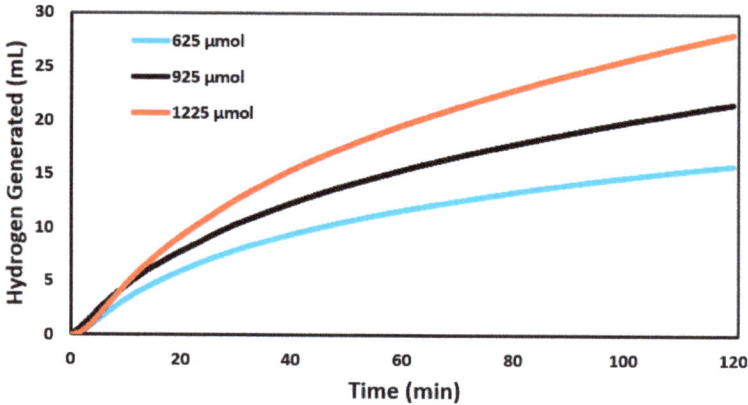

Figure 6. Volume of hydrogen generated over a period of two hours at varying doses of sodium borohydride.

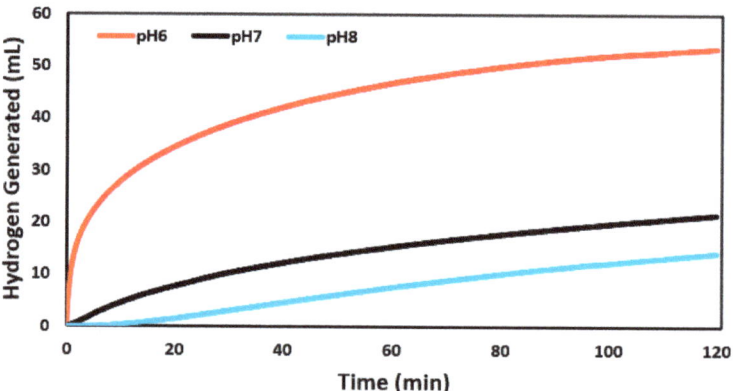

Figure 7. Volume of hydrogen generated over a period of two hours at varying pH solutions.

The AuFGLM was then tested at varying temperatures (Figure 8).

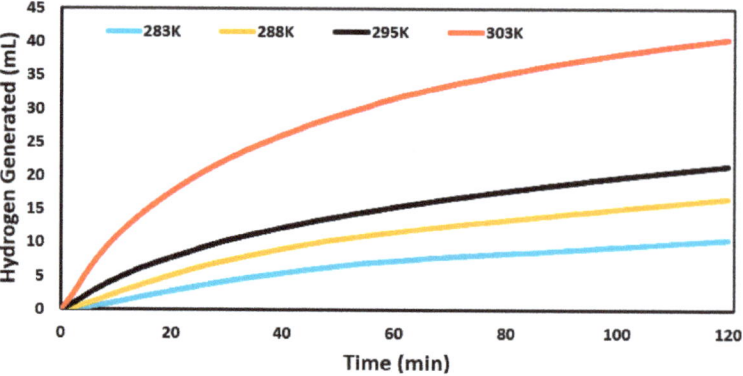

Figure 8. Volume of hydrogen generated over a period of two hours at varying solution temperatures.

For the temperature trial, the temperature of the reaction was cooled to 283 K and produced 10.6 mL of hydrogen gas at a rate of 0.0088 mL min^{-1}mg$_{cat}^{-1}$. At a temperature of 288 K, the reaction produced hydrogen gas at a rate of 0.0139 mL min^{-1} mg$_{cat}^{-1}$ and 16.7 mL after two hours. At room temperature (295 K), the reaction produced 21.6 mL of hydrogen gas at a rate of 0.0180 mL min^{-1}mg$_{cat}^{-1}$. Finally, when the temperature of the reaction was heated to 303 K an increase in hydrogen gas generation was observed, with 40.5 mL produced at a rate of 0.0337 mL min^{-1}mg$_{cat}^{-1}$. Based on these conditions it is clear that there is a positive linear relationship, with increases in temperature resulting in increases in hydrogen gas generation. According to Le Chatlier's principle, this means that the reaction is endothermic.

Once the temperature trials had been completed, data could be plugged into the Arrhenius Equation (2) to find the rate constant (k) at each temperature. In the following equation the variables are represented as the following; k is the rate constant of the reaction at the tested temperature (T) in Kelvin. The variable A represents the pre-exponential factor, and the variable R is the universal gas constant. Lastly, Ea represents the activation energy of the reaction.

$$k = Ae^{-\frac{Ea}{RT}} \qquad (2)$$

Using the natural log of the rate constant found at each temperature vs. that temperature divided by 1000 (Figure 9), the slope of the line allowed for the activation energy of AuFGLM to be determined to be 45.5 kJ mol^{-1}.

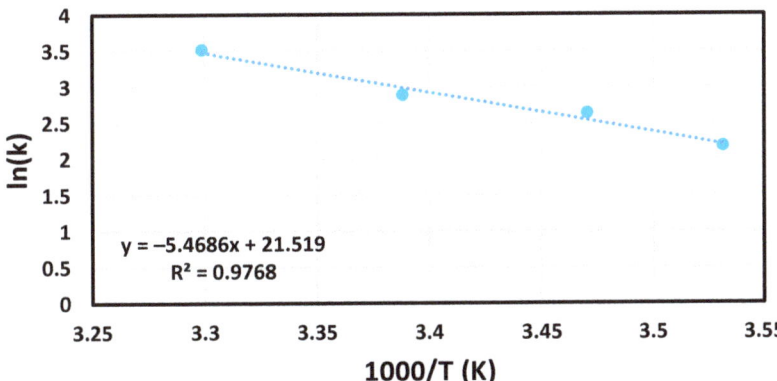

Figure 9. Arrhenius plot created by the using the Arrhenius Equation (3) and the temperature of the trials. The blue dot line indicated the trendline.

The activation energy of this catalyst was then compared to similar catalysts for the hydrolysis of NaBH$_4$, as seen in Table 1.

As shown in Table 1, AuFGLM has an activation energy that is competitive compared to other catalysts for the hydrolysis reaction of sodium borohydride. Nanoporous graphene oxide (PGO), a material that is similar to our unsupported FGLM, showed no ability to produce hydrogen in this reaction [42]. As such, FGLM on its own was not believed to be able to produce hydrogen on its own. When compared to bulk non-precious metal catalysts such as nickel and cobalt, AuFGLM had a significantly lower activation energy. Compared to unsupported gold nanoparticles, the AuFGLM again had a lower activation energy, indicating that the addition of the fused graphene-like material support aided in the gold nanoparticle's catalytic ability. Unsupported platinum nanoparticles, however, outperformed the composite, possibly hinting at platinum being a better catalyst than gold; however, this is just speculation, and more work needs to be done to confirm this. Other carbon-supported catalysts either performed better or worse than the AuFGLM, depending on the metal, again indicating that more research is needed to improve our understanding of which catalysts are best for this reaction.

Table 1. Comparison of activation energies.

Catalyst	E_a (kJ mol^{-1})	Temperature (K)	Reference
Ni	71	273–308	[33]
Raney-Nickel	63	273–308	[33]
Co	75	273–308	[33]
Ni–Co–B	62	281–301	[36]
Ru/ZIF	36.4	303–333	[37]
Ru-on-carbon	66.9	298–358	[38]
Pt/MWCNTs	46.2	283–303	[34]
Ag/MWCNTs	44.5	273–303	[12]
Pd/MWCNTs	62.7	273–303	[14]
Au/MWCNTs	21.1	273–303	[18]
PtNPs	39.2	283–303	[13]
BCD-AuNP	54.7	283–303	[11]
Pd Nanocup	58.9	283–303	[39]
AgNPs	50.3	273–303	[40]
PtNPs	50.3	273–303	[41]
PGO	N/A	298	[42]
AuFGLM	45.5	283–303	This Work

The final catalytic study conducted concerned the reusability of the catalyst. A trial was set up with standard conditions (295 K, pH 7, and a NaBH$_4$ dose of 925 µmol) that ran for two hours. After the two-hour period was completed, an additional 925 µmol dose was added to the same reaction vessel, marking the start of the second trial. This was repeated for a total of five trials, producing an average volume of 25.4 mL per two-hour trial, or 28% based on the theoretical maximum of 90 mL (Figure 10). The data show that the catalyst produced similar amounts of hydrogen for three consecutive trials, after which there was an increase in hydrogen generation. The increasing trend of the hydrogen generation is related to the binding of BH_4^- and H^- species on the surface of gold nanoparticles, as shown in Equation (3). Over time, these bonds became hydrolyzed, as shown in Equation (4) and improved the electrostatic stabilization of the AuNPs surface, making them more active [43]. These results not only indicate that the catalyst is stable and can be reused, cutting costs and materials, but also could mean that the catalyst becomes more catalytically activated with multiple uses.

$$2 \text{ Surface-Au} + BH_4^- \leftrightarrow \text{Surface-Au-H} + \text{Surface-Au-BH}_3^- \tag{3}$$

$$\text{Surface-Au-H} + \text{Surface-Au-BH}_3^- + HOH \rightarrow \text{Surface-Au} + H_2 + [BH_2(OH)]^- \tag{4}$$

Scheme 2 depicts one possible mechanism for the catalytic hydrolysis of NaBH$_4$ by AuFGLM. The gold nanoparticles are supported on the surface of the fused graphene-like material. In the solution, the NaBH$_4$ dissociated into borohydride ions $(BH_4)^-$ which then bonded with the nanoparticles. Water in the solution then attacked the boron in the borohydride and split off one hydrogen atom. This hydrogen atom bonded with a hydrogen atom from the BH$_4$, which then released itself as a diatomic hydrogen molecule. The remaining hydroxyl (OH) group from the water stayed fixed to the boron atom. This happened four times in total, at which point the resulting tetrahydroxyborate $[B(OH)_4^-]$ molecule left the nanoparticle and another borohydride ion took its place to restart the cycle.

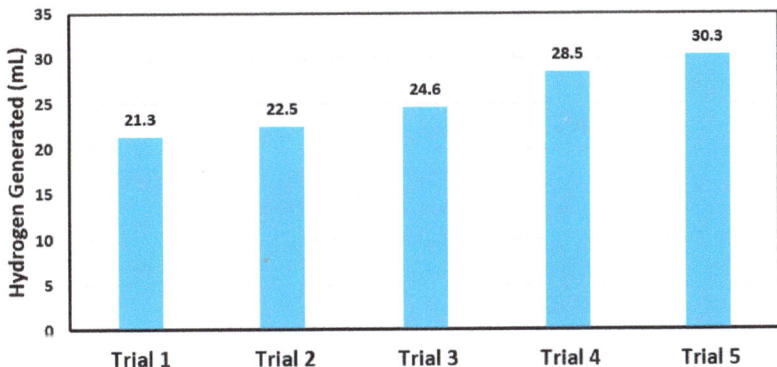

Figure 10. Volume of hydrogen gas produced for five consecutive two-hour trials in standard conditions.

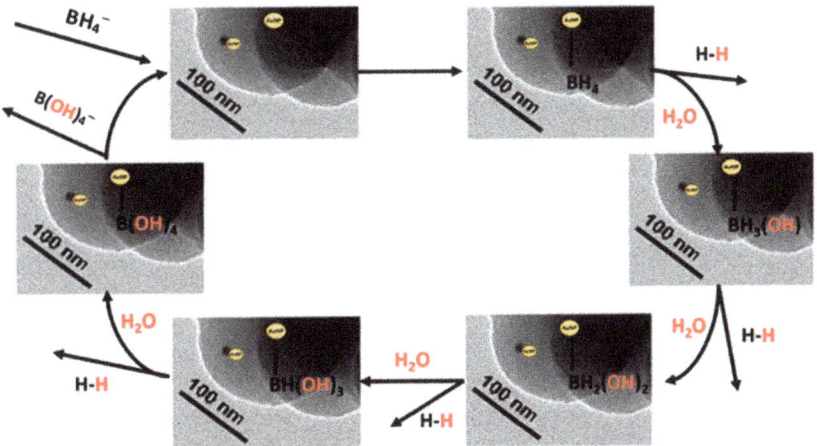

Scheme 2. Proposed mechanism for the catalytic hydrolysis of NaBH$_4$.

4. Conclusions

Fused graphene-like materials (FGLMs) have a structure that combines multiple graphene-like layers into a single cohesive material. The specific combination of different graphene-like layers can be tailored to achieve desired functionalities as a catalyst support. The FGLMs supporting the gold nanoparticles were characterized via TEM in order to determine the general morphology of the composite and the adhesive properties of the nanoparticles. The catalytic properties of the AuFGLM composite were tested via a hydrogen evolution reaction involving NaBH$_4$, where it was found that the optimal conditions for AuFGLM to produce hydrogen are increased temperatures, increased NaBH$_4$ doses, and decreased pH's. This catalyst produced the most hydrogen at a pH value of 6, at a rate of 0.0446 mL min^{-1}mg$_{cat}^{-1}$. Temperature data allowed the activation energy of this reaction as catalyzed by the AuFGLM to be 45.5 kJ mol^{-1}, which makes it competitive compared to other similar catalysts for this reaction. Additionally, it was found that the catalyst was stable for up to at least five consecutive trials, producing an average hydrogen volume of 25.4 mL. This stability, along with a competitive activation energy, makes this catalyst a suitable option for optimizing the hydrolysis of NaBH$_4$ for the generation of hydrogen as an alternative fuel source.

Author Contributions: E.B.: data curation, formal analysis, writing—original draft. Q.Q.: data curation, formal analysis, writing—original draft. T.M.A.-F.: conceptualization, validation, formal analysis, investigation, resources, supervision, writing—review & editing. All authors have read and agreed to the published version of the manuscript.

Funding: This research received no external funding.

Institutional Review Board Statement: Not applicable.

Informed Consent Statement: Not applicable.

Data Availability Statement: The data presented in this study are available on request from the corresponding author.

Acknowledgments: The corresponding author acknowledges Lawrence J. Sacks Professorship in Chemistry.

Conflicts of Interest: The authors declare no conflict of interest.

References

1. Ritchie, H.; Roser, M. Energy—Our World in Data. 2014. Available online: https://ourworldindata.org/energy (accessed on 15 April 2021).
2. Rodhe, H.A. Comparison of the Contribution of Various Gases to the Greenhouse Effect. *Science* **1990**, *248*, 1217–1219. [CrossRef] [PubMed]
3. Schlesinger, H.I.; Brown, H.C.; Finholt, E.; Gilbreath, J.R.; Hoekstra, H.R.; Hyde, E.K. Sodium Borohydride, Its Hydrolysis and its Use as a Reducing Agent and in the Generation of Hydrogen. *J. Am. Chem. Soc.* **1953**, *75*, 215–219. [CrossRef]
4. Biehler, E.; Whiteman, R.; Lin, P.; Zhang, K.; Baumgart, H.; Abdel-Fattah, T.M. Controlled Synthesis of ZnO Nanorods Using Different Seed Layers. *ECS J. Solid State Sci. Technol.* **2020**, *9*, 121008. [CrossRef]
5. Quach, Q.; Abdel-Fattah, T.M. Silver Nanoparticles functionalized Nanoporous Silica Nanoparticle grown over Graphene Oxide for enhancing Antibacterial effect. *Nanomaterials* **2022**, *12*, 3341. [CrossRef]
6. Dushatinski, T.; Abdel-Fattah, T.M. Carbon Nanotube Composite Mesh Film with Tunable Optoelectronic Performance. *ECS J. Solid State Sci. Technol.* **2015**, *4*, M1–M5. [CrossRef]
7. Abdel-Fattah, T.M.; Wixtrom, A.; Zhang, k.; Baumgart, H. Highly Uniform Self-Assembled Gold Nanoparticles over High Surface Area Dense ZnO Nanorod Arrays as Novel Surface Catalysts. *ECS J. Solid State Sci. Technol.* **2014**, *3*, M61–M64. [CrossRef]
8. Abdel-Fattah, T.M.; Wixtrom, A. Catalytic Reduction of 4-Nitrophenol Using Gold Nanoparticles Supported On Carbon Nanotubes. *ECS J. Solid State Sci. Technol.* **2014**, *3*, M18–M20. [CrossRef]
9. Dushatinski, T.; Huff, C.; Abdel-Fattah, T.M. Characterization of electrochemically deposited films from aqueous and ionic liquid cobalt precursors toward hydrogen evolution reactions. *Appl. Surf. Sci.* **2016**, *385*, 282–288. [CrossRef]
10. Huff, C.; Dushatinski, T.; Barzanjii, A.; Abdel-Fattah, N.; Barzanjii, K.; Abdel-Fattah, T.M. Pretreatment of gold nanoparticle multi-walled carbon nanotube composites for catalytic activity toward hydrogen generation reaction. *ECS J. Solid State Sci. Technol.* **2017**, *6*, M69–M71. [CrossRef]
11. Quach, Q.; Biehler, E.; Elzamzami, A.; Huff, C.; Long, J.M.; Abdel-Fattah, T.M. Catalytic Activity of Beta-Cyclodextrin-Gold Nanoparticles Network in Hydrogen Evolution Reaction. *Catalysts* **2021**, *11*, 118. [CrossRef]
12. Huff, C.; Long, J.M.; Aboulatta, A.; Heyman, A.; Abdel-Fattah, T.M. Silver Nanoparticle/Multi-Walled Carbon Nanotube Composite as Catalyst for Hydrogen Production. *ECS J. Solid State Sci. Technol.* **2017**, *6*, M115–M118. [CrossRef]
13. Biehler, E.; Quach, Q.; Abdel-Fattah, T.M. Silver Nanoparticle-Decorated Fused Carbon Sphere Composite as a Catalyst for Hydrogen Generation. *Energies* **2023**, *16*, 5053. [CrossRef]
14. Huff, C.; Long, J.M.; Heyman, A.; Abdel-Fattah, T.M. Palladium Nanoparticle Multiwalled Carbon Nanotube Composite as Catalyst for Hydrogen Production by the Hydrolysis of Sodium Borohydride. *ACS Appl. Energy Mater.* **2018**, *1*, 4635–4640. [CrossRef]
15. Sengani, M.; Grumezescu, A.M.; Rajeswari, V.D. Recent Trends and Methodologies in Gold Nanoparticle Synthesis—A Prospective Review on Drug Delivery Aspect. *OpenNano* **2017**, *2*, 37–46. [CrossRef]
16. Upadhyayula, V.K.K. Functionalized Gold Nanoparticle Supported Sensory Mechanisms Applied in Detection of Chemical and Biological Threat Agents: A Review. *Anal. Chim. Acta* **2012**, *715*, 1–18. [CrossRef]
17. Elahi, N.; Kamali, M.; Baghersad, M.H. Recent Biomedical Applications of Gold Nanoparticles: A Review. *Talanta* **2018**, *184*, 537–556. [CrossRef]
18. Huff, C.; Dushatinski, T.; Abdel-Fattah, T.M. Gold Nanoparticle/Multi-Walled Carbon Nanotube Composite as Novel Catalyst for Hydrogen Evolution Reactions. *Int. J. Hydrogen Energy* **2017**, *42*, 18985–18990. [CrossRef]
19. Mitsudome, T.; Kaneda, K. Gold nanoparticle Catalysts for Selective Hydrogenations. *Green Chem.* **2013**, *15*, 2636–2654. [CrossRef]
20. Zou, X.; Xu, W.; Liu, G.; Panda, D.; Chen, P. Size-Dependent Catalytic Activity and Dynamics of Gold Nanoparticles at the Single-Molecule Level. Size-Dependent Catalytic Activity and Dynamics of Gold Nanoparticles at the Single-Molecule Level. *J. Am. Chem. Soc.* **2010**, *132*, 138–146. [CrossRef]

21. Antolini, E. Formation, Microstructural Characteristics and Stability of Carbon Supported Platinum Catalysts for Low Temperature Fuel Cells. *J. Mater. Sci.* **2003**, *38*, 2995–3005. [CrossRef]
22. Osborne, J.; Horten, M.R.; Abdel-Fattah, T.M. Gold Nanoparticles Supported Over Low-Cost Supports for Hydrogen Generation from a Hydrogen Feedstock Material. *ECS J. Solid State Sci. Technol.* **2020**, *9*, 071004. [CrossRef]
23. Baig, N.; Kammakakam, I.; Falath, W. Nanomaterials: A review of synthesis methods, properties, recent progress, and challenges. *Mater. Adv.* **2021**, *2*, 1821–1871. [CrossRef]
24. Ivanovskii, A.L. Graphene-Based and Graphene-Like Materials. *Russ. Chem. Rev.* **2012**, *81*, 571–605. [CrossRef]
25. Li, K.; Liu, Q.; Cheng, H.; Hu, M.; Zhang, S. Classification and carbon structural transformation from anthracite to natural coaly graphite by XRD, Raman spectroscopy, and HRTEM. *Spectrochim. Acta Part A Mol. Biomol. Spectrosc.* **2020**, *249*, 119286. [CrossRef]
26. Ghosh, S.; Calizo, I.; Teweldebrhan, D.; Pokatilov, E.P.; Nika, D.L.; Balandin, A.A.; Bao, W.; Miao, F.; Lau, C.N. Extremely high thermal conductivity of graphene: Prospects for thermal management applications in nanoelectronic circuits. *Appl. Phys. Lett.* **2008**, *92*, 151911. [CrossRef]
27. Fadeel, B.; Bussy, C.; Merino, S.; Vázquez, E.; Flahaut, E.; Mouchet, F.; Evariste, L.; Gauthier, L.; Koivisto, A.J.; Vogel, U.; et al. Safety Assessment of Graphene-Based Materials: Focus on Human Health and the Environment. *ACS Nano.* **2018**, *12*, 10582–10620. [CrossRef]
28. Krishnamurthy, S.; Esterle, A.; Sharma, N.C.; Sahi, S.V. Yucca-derived synthesis of gold nanomaterial and their catalytic potential. *Nanoscale Res. Lett.* **2014**, *9*, 627. [CrossRef]
29. Perumbilavil, S.; Sankar, P.; Rose, T.P.; Philip, R. White light Z-scan measurements of ultrafast optical nonlinearity in reduced graphene oxide nanosheets in the 400–700 nm region. *Appl. Phys. Lett.* **2015**, *107*, 051104. [CrossRef]
30. Lee, A.Y.; Yang, K.; Anh, N.D.; Park, C.; Lee, S.M.; Lee, T.G.; Jeong, M.S. Raman study of D* band in graphene oxide and its correlation with reduction. *Appl. Surf. Sci.* **2021**, *536*, 147990. [CrossRef]
31. Govindaraju, S.; Ramasamy, M.; Baskaran, R.; Ahn, S.J.; Yun, K. Ultraviolet light and laser irradiation enhances the antibacterial activity of glucosamine-functionalized gold nanoparticles. *Int. J. Nanomed.* **2015**, *10*, 67–78. [CrossRef]
32. Lai, C.H.; Wang, G.A.; Ling, T.K.; Wang, T.J.; Chiu, P.K.; Chau, Y.F.C.; Huang, C.C.; Chian, F.P. Near infrared surface-enhanced Raman scattering based on star-shaped gold/silver nanoparticles and hyperbolic metamaterial. *Sci. Rep.* **2017**, *7*, 5446. [CrossRef] [PubMed]
33. Kaufman, C.M. Hydrogen generation by hydrolysis of sodium tetrahydroborate: Effects of acids and transition metals and their salts. *J. Chem. Soc. Dalton Trans.* **1985**, *2*, 307–313. [CrossRef]
34. Huff, C.; Quach, Q.; Long, J.M.; Abdel-Fattah, T.M. Nanocomposite Catalyst Derived from Ultrafine Platinum Nanoparticles and Carbon Nanotubes for Hydrogen Generation. *ECS J. Solid State Sci. Technol.* **2020**, *9*, 101008. [CrossRef]
35. Grzeschik, R.; Schäfer, D.; Holtum, T.; Küpper, S.; Hoffmann, A.; Schlücker, S. On the Overlooked Critical Role of the pH Value on the Kinetics of the 4-Nitrophenol NaBH4-Reduction Catalyzed by Noble Metal Nanoparticles (Pt, Pd, Au). *J. Phys. Chem. C.* **2020**, *124*, 2939–2944. [CrossRef]
36. Ingersoll, J.C.; Mani, N.; Thenmozhiyal, J.C.; Muthaiah, A. Catalytic hydrolysis of sodium borohydride by a novel nickel–cobalt–boride catalyst. *J. Power Sources* **2007**, *173*, 450–457. [CrossRef]
37. Tuan, D.D.; Lin, K.-Y.A. Ruthenium supported on ZIF-67 as an enhanced catalyst for hydrogen generation from hydrolysis of sodium borohydride. *Chem. Eng. J.* **2018**, *351*, 48–55. [CrossRef]
38. Zhang, J.S.; Delgass, W.N.; Fisher, T.S.; Gore, J.P. Kinetics of Ru-catalyzed sodium borohydride hydrolysis. *J. Power Sources* **2007**, *164*, 772–781. [CrossRef]
39. Biehler, E.; Quach, Q.; Huff, C.; Abdel-Fattah, T.M. Organo-Nanocups Assist the Formation of Ultra-Small Palladium Nanoparticle Catalysts for Hydrogen Evolution Reaction. *Materials* **2022**, *15*, 2692. [CrossRef]
40. Huff, C.; Long, J.M.; Abdel-Fattah, T.M. Beta-Cyclodextrin Assisted Synthesis of Silver Nanoparticle Network and its Application in a Hydrogen Generation Reaction. *Catalysts* **2020**, *10*, 1014. [CrossRef]
41. Huff, C.; Biehler, E.; Quach, Q.; Long, J.M.; Abdel-Fattah, T.M. Synthesis of highly dispersive platinum nanoparticles and their application in a hydrogen generation reaction. *Colloids Surf. A Physicochem. Eng. Asp.* **2021**, *610*, 125734. [CrossRef]
42. Zhang, H.; Feng, X.; Cheng, L.; Hou, X.; Li, Y.; Han, S. Non-noble Co anchored on nanoporous graphene oxide, as an efficient and long-life catalyst for hydrogen generation from sodium borohydride. *Colloids Surf. A Physicochem. Eng. Asp.* **2018**, *563*, 112–119. [CrossRef]
43. Deraedt, C.; Salmon, L.; Gatard, S.; Ciganda, R.; Hernandez, E.; Ruiz, J.; Astruc, D. Sodium borohydride stabilizes very active gold nanoparticle catalysts. *Chem. Commun.* **2014**, *50*, 14194–14196. [CrossRef] [PubMed]

Disclaimer/Publisher's Note: The statements, opinions and data contained in all publications are solely those of the individual author(s) and contributor(s) and not of MDPI and/or the editor(s). MDPI and/or the editor(s) disclaim responsibility for any injury to people or property resulting from any ideas, methods, instructions or products referred to in the content.

Article

Effect of Graphene Oxide on the Mechanical Properties and Durability of High-Strength Lightweight Concrete Containing Shale Ceramsite

Xiaojiang Hong [1,2], Jin Chai Lee [1,*], Jing Lin Ng [1,2], Zeety Md Yusof [3], Qian He [1,2] and Qiansha Li [1,2]

[1] Department of Civil Engineering, Faculty of Engineering, Technology and Built Environment, UCSI University, Kuala Lumpur 56000, Malaysia
[2] Department of Civil Engineering, Faculty of Civil and Hydraulic Engineering, Xichang University, Xichang 615013, China
[3] Department of Civil Engineering, Faculty of Civil Engineering and Built Environment, Universiti Tun Hussein Onn Malaysia, Parit Raja 86400, Malaysia
* Correspondence: leejc@ucsiuniversity.edu.my or jinlee861@gmail.com

Abstract: An effective pathway to achieve the sustainable development of resources and environmental protection is to utilize shale ceramsite (SC), which is processed from shale spoil to produce high-strength lightweight concrete (HSLWC). Furthermore, the urgent demand for better performance of HSLWC has stimulated active research on graphene oxide (GO) in strengthening mechanical properties and durability. This study was an effort to investigate the effect of different contents of GO on HSLWC manufactured from SC. For this purpose, six mixtures containing GO in the range of 0–0.08% (by weight of cement) were systematically designed to test the mechanical properties (compressive strength, flexural strength, and splitting tensile strength), durability (chloride penetration resistance, freezing–thawing resistance, and sulfate attack resistance), and microstructure. The experimental results showed that the optimum amount of 0.05% GO can maximize the compressive strength, flexural strength, and splitting tensile strength by 20.1%, 34.3%, and 24.2%, respectively, and exhibited excellent chloride penetration resistance, freezing–thawing resistance, and sulfate attack resistance. Note that when the addition of GO was relatively high, the performance improvement in HSLWC as attenuated instead. Therefore, based on the comprehensive analysis of microstructure, the optimal addition level of GO to achieve the best mechanical properties and durability of HSLWC is considered to be 0.05%. These findings can provide a new method for the use of SC in engineering.

Keywords: HSLWC; mechanical properties; durability; microstructure; GO

1. Introduction

Lightweight concrete (LWC), compared with normal-weight concrete (NWC), has been widely and prosperously developed in the past decades due to its lower density and higher thermal insulation performance [1]. LWC can reduce the weight of cement by up to 20% without affecting the required strength, thus contributing to saving raw materials and transportation costs [2]. In addition, it was reported that LWC could reduce thermal energy consumption by approximately 15% in order to achieve thermal comfort in European buildings [3]. At present, many researchers have further explored the potential of mechanical properties and durability of LWC in order to develop high-strength lightweight concrete (HSLWC) with excellent performance [4,5]. Aslam et al. found that using oil-palm-boiler clinker (OPBC) to replace different proportions of coarse aggregate produced lightweight concrete with a 28-day compressive strength of 47 MPa [6]. Kılıç et al. showed that using basalt pumice produced structural HSLWC with a 90-day compressive strength of 43.8 MPa [7]. Rossignolo et al. reported that using local lightweight aggregates in Brazil as coarse aggregates could manufacture HSLWC with a 28-day compressive strength of 53.6 MPa [8]. As the scope of application expands, HSLWC has gradually evolved from nonstructural materials to structural materials. Typically, HSLWC can be used as an efficient structural material to increase the number of stories in high-rise buildings, extend the span of bridges, and strengthen the corrosion resistance of offshore platforms [9,10]. Furthermore, many countries are investigating the production of HSLWC from a variety of construction waste and recycled aggregate, such as ceramicite, fly ash, OPBC, pumice stone, and geopolymers, which has been facilitated in terms of environmental protection, the economy, and sustainable development [11–13].

Natural resources such as river sand and stone have been excessively consumed, while the accumulation of large quantities of shale spoil has also produced ecological pollution in China. Therefore, the government requires considerable human and material resources for the disposal and recycling of shale spoil every year. Under the incentive of the increasing demand of the light aggregate industry and the updating of ceramsite production technology, shale spoil can be mass produced into shale ceramsite (SC) and shale pottery sand (SPS) with different particle size distribution sunder high-temperature calcination processing [14]. Based on the characteristics of an increased number of pores, lighter weight, and higher strength, SC is considered as a good lightweight aggregate for producing HSLWC [15]. On the one hand, the bond strength of the interface between cement slurry and SC observed by SEM is an important contribution to strengthening the mechanical properties and durability of HSLWC. On the other hand, the porous structure of SC is not conducive to the compactness of concrete, thus limiting the physical reinforcement ability. With the same ratio of sand, the higher the content of SC, the lower the ultimate compressive strength of LWAC [16]. Although adversely affected by the high water absorption of SC, the slump of HSLWC is satisfactory under the condition of a low water–binder ratio [17].

Increasing the compressive strength to 55 MPa and improving durability are challenging problems for HSLWC, which is fundamentally limited by the lightweight and porous characteristics of aggregate [18]. The use of steel fiber, carbon fiber, and resin fiber as additives to strengthen and toughen the aggregate is a common way to overcome these problems [19,20]. Adding an appropriate amount of nanosilica and fly ash resulted in a positive synergistic effect on the mechanical properties and durability of SC concrete [21]. Nevertheless, nanoscale pores and microcracks still nucleate and grow in the hydration reaction of cement mortar during long-term loading [22]. Graphene oxide (GO) is a unique two-dimensional nanosheet structure that has attracted great attention in the research to improve the performance of cement mortar and concrete [23,24]. At present, the main industrial GO preparation process involves extraction from graphene with the modified Hummer's method, and then cooling and drying in a vacuum [25]. With the advances in and development of preparation technology, GO is bound to be mass manufactured at a lower fabrication cost. Additionally, previous studies have demonstrated that the amount of GO added to cement-based materials has been optimized to a lower dosage [26,27]. Therefore, it is economically and technically feasible for the addition of GO to strengthen the performance of concrete. In general, the mechanism through which GO enhances concrete performance is currently in an intense exploration stage, and two strongly supportive theories are emphasized as follows: (1) the formation of flower-like crystals to regulate hydration reaction [28]; (2) the provision of nanoscale filling to enhance compactness [29].

The application of GO in cement-based materials mainly focuses on workability and mechanical properties. The better adhesion between GO and cement mortar leads to the lower slump and shorter setting time for fresh concrete [30]. GO mixed with in different proportions in cement mortar results in different degrees of enhancement in the mechanical properties. The addition of 0.03% GO as the optimal dose greatly reduced the total porosity of cement mortar and correspondingly increased the compressive strength by more than 40% [31]. Simultaneously, using the same proportion of GO was the most effective in promoting the growth and regulation of flower-like crystals [32]. When the addition amount exceeded 0.04%, the enhancement effect was inhibited due to the agglomeration of GO [33]. Similar conclusions on mechanical properties have been confirmed by other researchers for the application of pervious concrete and ultra-high-performance concrete (UHPC) [34,35]. However, limited information has been presented on the influence of GO on durability. Yu et al. reported that the incorporation of GO had a favorable impact on the durability of UHPC prepared from recycled sand [36]. Zeng et al. experimentally found that GO extended the life of cement mortars by more than two times when the specimens were exposed to sulfate attack [37]. In general, the many research achievements regarding GO have mainly concentrated on cement mortar and UHPC [38,39], but few studies have been conducted to investigate the influence of GO on the mechanical properties and durability of HSLWC. Hence, it is imperative to further analyze the potential of GO to stimulate HSLWC with SC as aggregate, so as to provide practical reference for construction.

The aims of this study were to use SC as a coarse aggregate to produce grade 60 HSLWC and to design six groups of GO mixtures with different contents to investigate the reinforcement effect and determine the optimal GO content. The following three indicators were used for comprehensive verification: (1) mechanical properties (compressive strength, flexural strength, and splitting tensile strength) tests in accordance with the GB/T 50081-2002 procedure; (2) durability (chloride penetration resistance, freezing–thawing resistance, and sulfate attack resistance) tests in accordance with the GB/T 50082-2009 procedure; (3) microstructure characterization.

2. Materials and Methods

2.1. Materials

Ordinary Portland cement (P.O 42.5 R), as the binder in all mixtures, was produced from Xichang Aerospace Co., Ltd. (Xichang, China). Its apparent density, specific surface area, and 28-day compressive strength were 3080 kg/m^3, 320 m^2/kg, and 47.4 MPa, respectively. All aggregates were manufactured by Hubei Huiteng Aggregate Co., Ltd. (Yichang, China). Their physical parameters are listed in Table 1, and their appearance is shown in Figure 1. In addition, the cylinder compression strength of SC was 5.2 MPa. Specifically, it can be seen from Figure 2a that SC presented a porous honeycomb structure on the surface, as measured by SEM. The fineness modulus of SPS was 2.96, which meets the requirements for medium sand in JGJ52-2016. GO, purchased from Suzhou Tanfeng Technology Graphene Technology Co., Ltd. (Suzhou, China), presented as a brownish powder, and its property parameters are illustrated in Table 2. Figure 2b shows that GO was a nanoscale sheet with a typical fine and dense wrinkle morphology, as measured by SEM. Fourier infrared spectroscopy (FTIR) of GO is shown in Figure 3. The four stretching vibration peaks of 3200, 1724, 1617, and 1064 cm^{-1} confirmed the activity of the typical oxidation functional groups -OH, C=O, C=C, and C-O, respectively. These functional groups play an important role in improving the hydrophilicity of GO, thereby contributing to better dispersion in mixtures [40]. Grade 1 of fly ash (FA) with a specific surface area of 420 m^2/kg and loss on ignition of 3.48% was selected as the mineral admixture. The polycarboxylate superplasticizer (PS) not only had a water reduction rate of 12% but was also used as an active agent for dispersing GO.

Table 1. Physical properties of the aggregates.

Aggregate	Type	Density Rank (kg/m^3)	Apparent Density (kg/m^3)	Particle Size (mm)	Water Absorption (24 h) (%)
SC	coarse	800	1425	5–15	4.6
SPS	fine	700	1638	0–3	1.36

Figure 1. The images of aggregates: (**a**) SC; (**b**) SPS.

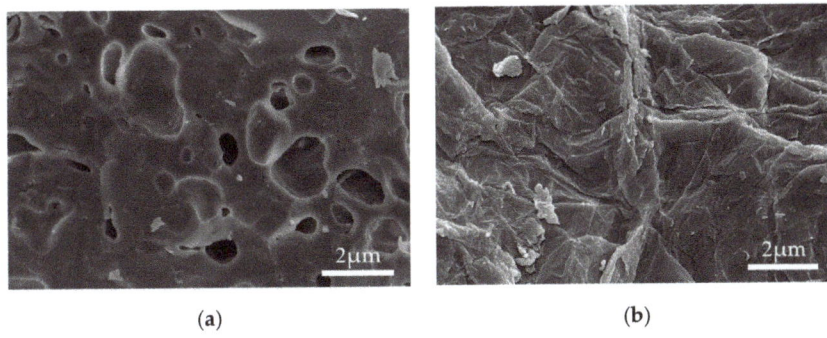

Figure 2. SEM images: (**a**) SC; (**b**) GO.

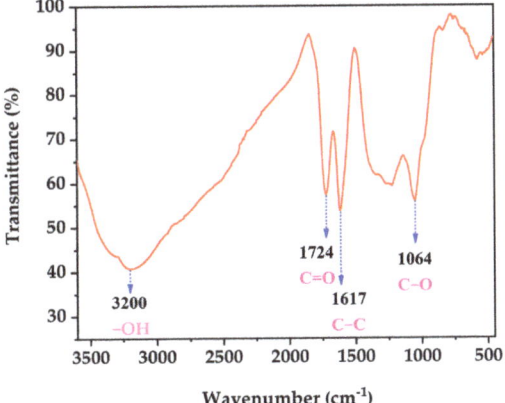

Figure 3. FTIR spectra of GO.

Table 2. Property parameters of GO.

Specific Surface Area (m^2/g)	Layers	Thickness (nm)	Diameter (μm)	Purity (%)	Oxygen Content (%)	Carbon Content (%)
100–300	1–2	~1	10–30	>95	>33	>66

2.2. Mix Proportions

The control mix proportion (GO-0) with a high-strength grade of 60 was calculated according to the absolute volume method according to GB/T 31387-2015. The advantage of this method is that the slurry plays the core role in improving strength, and the lightweight aggregate plays an auxiliary role of filling and reducing the apparent density. The other five trial mix proportions, numbered as GO-2, GO-4, GO-5, GO-6, and GO-8, were prepared, in which the added GO content was 0.02%, 0.04%, 0.05%, 0.06%, and 0.08% (by weight of cement), respectively. According to the prior studies on replacing cementitious materials with mineral admixtures to further obtain better fluidity and durability [41], the FA used in this study had a reasonable setting of 20%. The mix proportion of the six mixes is listed in Table 3.

Table 3. Mix proportion (kg/m^3).

No.	Cement	Water	SC	SPS	FA	PS	GO
GO-0	440	170	380	380	110	11	0
GO-2	440	170	380	380	110	11	0.088
GO-4	440	170	380	380	110	11	0.176
GO-5	440	170	380	380	110	11	0.220
GO-6	440	170	380	380	110	11	0.264
GO-8	440	170	380	380	110	11	0.352

2.3. Preparation and Curing

According to our previous research [42], PS needs to be used as an active agent with GO to achieve uniform dispersion in water under ultrasonic vibration for 30 min to finally prepare GO solution (GOS). GOS dispersion appeared uneven and particles were suspended when the GO content exceeded 0.05%. It was recommended that the prepared GO be used immediately. Furthermore, SC was presaturated for 24 h before casting, followed by naturally drying its surface in open air for reducing water absorption of SC. The mixing procedure was as follows: First, put the SC and SPS into a paddle mixer and mix for three minutes. Second, add cement and FA and continue mixing for three minutes. Third, mix 70% GOS solution and stir for another three minutes. Finally, add the remaining GOS into the mixer and mix for two minutes until the composite is well blended.

Each specimen casting was loaded into the plastic mold and vibrated at low speed on a vibration table. The specimen wrapped with plastic film was demolded after curing at ambient temperature for 24 h, and then placed in a standard curing box for curing until the test age. All performance tests were carried out according to the curing regime after reaching the specified age. Detailed mixing procedures and experimental items are shown in Figure 4. The results of all properties are reported as the average of the three specimens.

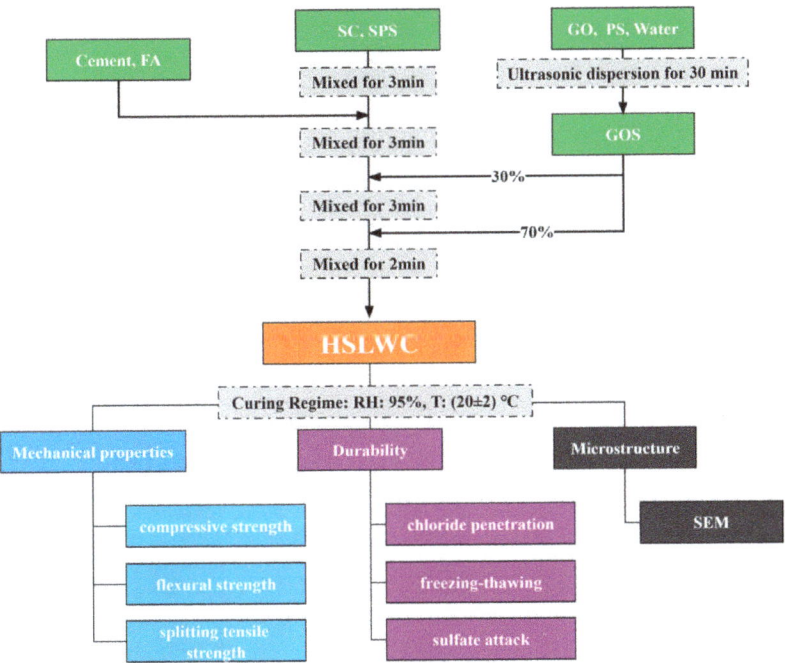

Figure 4. Mixing procedure and experimental items of HSLWC.

2.4. Experimental Methodology

2.4.1. Mechanical Properties

In order to evaluate the effect of GO on the growth trend in compressive strength, the specimens with sizes of 100 × 100 × 100 mm were cast, at the ages of 1 d, 3 d, 7 d, 28 d, and 56 d for testing on a servo pressure testing machine (YAD-2015), as shown in Figure 5a. In addition, specimens with sizes of 100 × 100 × 400 mm were used to test the flexural strength at the age of 28d on a servo universal testing machine (RE-8030), as shown in Figure 5b. Specimens with sizes of 100 × 100 × 100 mm were used to test the splitting tensile strength at the age of 28 days on a computer control pressure testing machine (WAW-1000), as shown in Figure 5c.

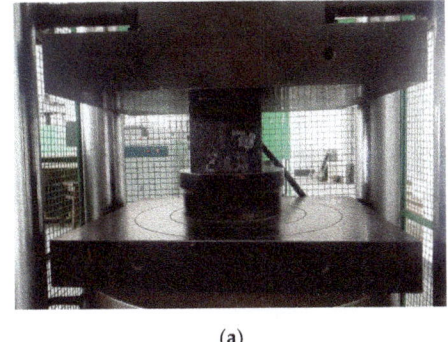

(a)

(b)

Figure 5. *Cont.*

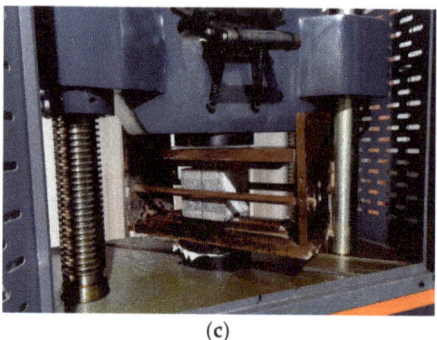

(c)

Figure 5. Test device for mechanical properties: (**a**) compressive strength test, (**b**) flexural strength test, and (**c**) splitting tensile strength test.

2.4.2. Durability

The rapid chloride ion migration coefficient (RCM) method was used to determine the chloride penetration resistance of HSLWC. The test pieces were cut equally from the standard cylinder specimen with a diameter of 100 mm and a height of 100 mm when the curing age reached 28 days, and then subjected to vacuum saturation treatment prior to testing. During the test, the test pieces were fixed on an RCM test device (RCM-NTB), as shown in Figure 6a. The voltage, average temperature, and duration were set to 30 V, 24 °C, and 24 h, respectively. Finally, the test piece, which was divided into two parts along the diameter direction, was used to measure the chloride intrusion depth on the cutting surface with a 0.1 mol/L $AgNO_3$ solution. The chloride ion migration coefficient (DRCM) (Formula S1) is a quantitative index used to evaluate the permeability resistance.

In this study, the quick-freezing method with water freezing and water melting as the test environment was adopted, and specimens with sizes of 100 × 100 × 400 mm were cast for testing every 25 cycles. Before cycling, the specimen was first incubated for 24 days in standard curing and then immersed in water with the temperature of (20 ± 2) °C for 4 days in freeze–thaw testing machine. The test was designed for a total of 250 cycles; each cycle took 4 h; at the highest and lowest temperatures were −18 °C and 5 °C, respectively. The mass loss rate (Formula S2) and relative elastic modulus (Formula S3) were specified as quantitative indices to evaluate the frost resistance of HSLWC. The freeze–thaw testing machine (JCD-40S) and dynamic elastic modulus testing device (NELD-DTV) are shown in Figure 6b,c.

Specimens with sizes of 100 × 100 × 100 mm were prepared to measure the resistance to sulfate attack under wet and dry cycles. After 26 days of standard curing, all specimens were moved to an oven with a temperature of 80 ± 5 °C to dry for 2 days. In principle, the volume content of 5% Na_2SO_4 aqueous solution should reach at least 20 mm to the top surface of the topmost specimens, which can provide a good erosion environment. The stages and time of each cycle were as follows: we placed the specimens in solution (10 min), soaked (15 h), discharged the solution (15 min), air dried (1 h), oven dried (6 h), and then cooled the specimen (2 h). The test data were collected once every 30 cycles, for a total of 150 cycles. The mass loss rate (Formula S4) and corrosion resistance coefficient (Formula S5) were used as quantitative indicators of the sulfate resistance of HSLWC. The sulfate attack testing machine (NELD-VS830) is shown in Figure 6d.

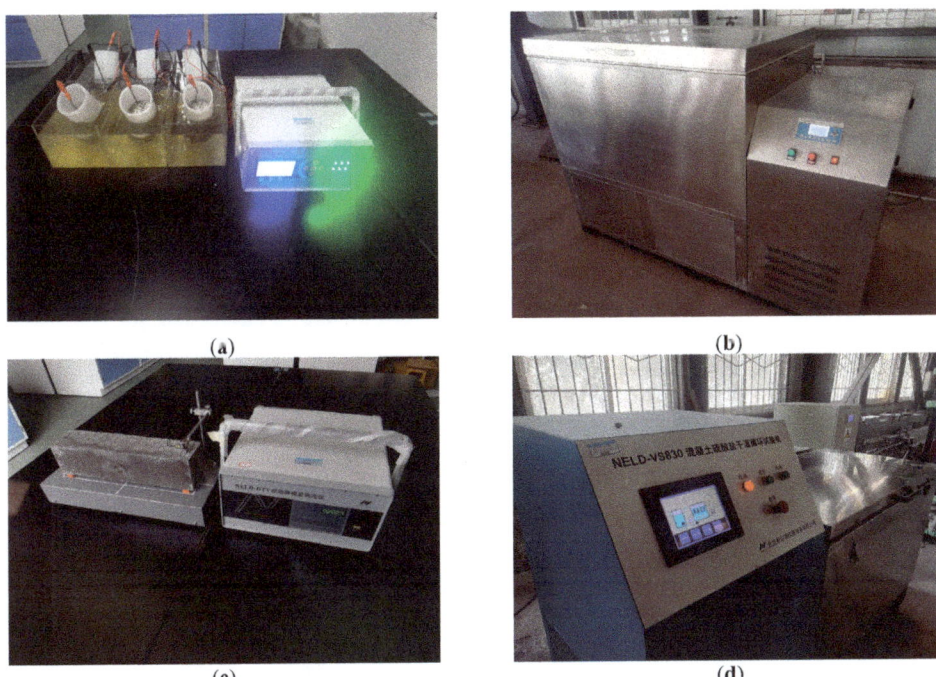

Figure 6. Test device for durability: (**a**) test device for rapid chloride ions migration (RCM) method, (**b**) freezing–thawing testing machine, (**c**) test device for dynamic modulus of elasticity, and (**d**) sulfate attack testing machine.

2.4.3. Microstructure

It was necessary to accurately analyze the mechanism of GO application in HSLWC. The microscopic characteristics of blocks with sizes of approximately $10 \times 10 \times 5$ mm, which were taken from each group of mix proportion specimens at an age of 28 days, were observed from through SEM after gold-spray treatment.

3. Results and Discussion

3.1. Mechanical Properties

3.1.1. Compressive Strength

Table 4 lists in detail the oven-dry density and compressive strength of specimens with different GO contents at 28 days under standard curing. All specimens had an oven-dry density in the range of 1696–1728 kg/m^3 and a compressive strength in the range of 61.88–74.32 MPa, whereas the classic HSLWC has a density of no more than 1850 kg/m^3 and a compressive strength of 35–79 MPa according to Monteiro et al. [43]. The specimens with GO (GO-5) obtained the highest compressive strength of 74.32 MPa, presenting a maximum increase of 20.1% compared with that of those without GO (GO-0). In addition, the specific strength (C/D in Table 4), defined as the ratio of compressive strength to density, is an important characteristic in assessing structural performance. The specimens without GO obtained a specific strength of 36.5 kN·m/kg, while the specimens with a GO content of 0.05% obtained a higher specific strength of 43.3 kN·m/kg. Shafigh et al. reported that the LWC produced with OPBC as coarse aggregate had a specific strength of 30.9 kN·m/kg [44]. Evangelista et al. found that HSLWC containing expanded shale had a specific strength of 36.3 kN·m/kg [45].

Table 4. The results and analysis of density and mechanical properties.

Mix No.	Density (kg/m³)	Compressive Strength (MPa)	C/D (kN·m/kg)	Ratio (%) F/C	Ratio (%) S/C
GO-0	1696	61.88	36.5	6.5	6.8
GO-2	1705	64.87	38.0	6.7	6.2
GO-4	1712	71.22	41.6	6.1	6.9
GO-5	1715	74.32	43.3	6.2	7.0
GO-6	1719	72.74	42.3	6.0	7.0
GO-8	1728	69.26	40.1	5.9	7.1

Figure 7 illustrates the compressive strength results of specimens with different GO incorporation contents at 1, 3, 7, 28, and 56 days under standard curing. Evidently, the specimens incorporating GO had a higher compressive strength than those without GO at each age, indicating that GO produced a positive and efficient contribution. However, the compressive strength showed a nonlinear trend of first increasing and then decreasing with the increase in GO content, suggesting that there is an optimal amount of GO to achieve maximum strength. It was noted that the optimum amount in this study was 0.05% of the mix (GO-5). Similar conclusions were drawn from the application of GO in UHPC according to Chu et al. and Yu. et al. [34,36]. The 3-day compressive strengths of GO-4, GO-5, GO-6, and GO-8 were significantly higher than the 7-day compressive strength of GO-0, and the 7-day compressive strength of GO-5 was very close to the 28-day compressive strength of GO-0, indicating that GO accelerated strength formation and thus shortened the curing time. This advantage provides economic convenience for the acceleration of template turnover and the reduction in construction costs [46].

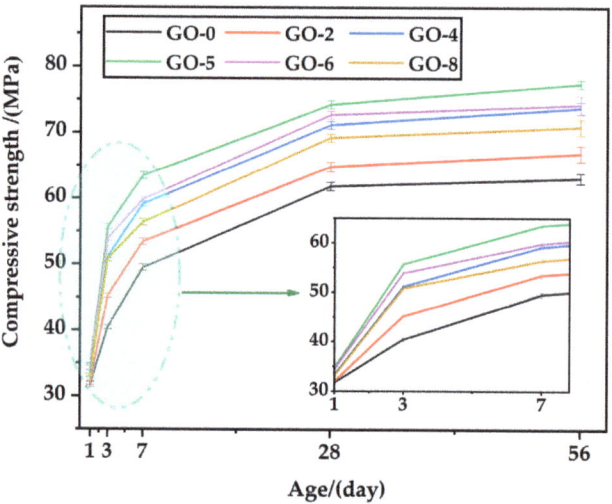

Figure 7. Compressive strength of HSLWC at different ages.

3.1.2. Flexural Strength

Figure 8a shows the flexural strength results of specimens with different GO incorporation contents at 28 days under standard curing. It was found that the flexural strength of the specimens incorporating GO was significantly higher than those without GO, which was attributed to the enhancement provided by GO. With the increase in GO content from 0 to 0.08%, the flexural strength presented a strong parabola trend. The flexural strengths of GO-0, GO-2, GO-4, GO-5, GO-6, and GO-8 were 6.47, 7.23, 8.11, 8.69, 8.34, and 8.25 MPa, respectively. Compared with GO-0, the flexural strengths of GO-2, GO-4, GO-5, GO-6, and GO-8 increased by 11.7%, 25.3%, 34.3%, 28.9%, and 27.4%, respectively. This result indicated that the optimal content of GO added to HSLWC was 0.05% in terms of improving the flexural strength.

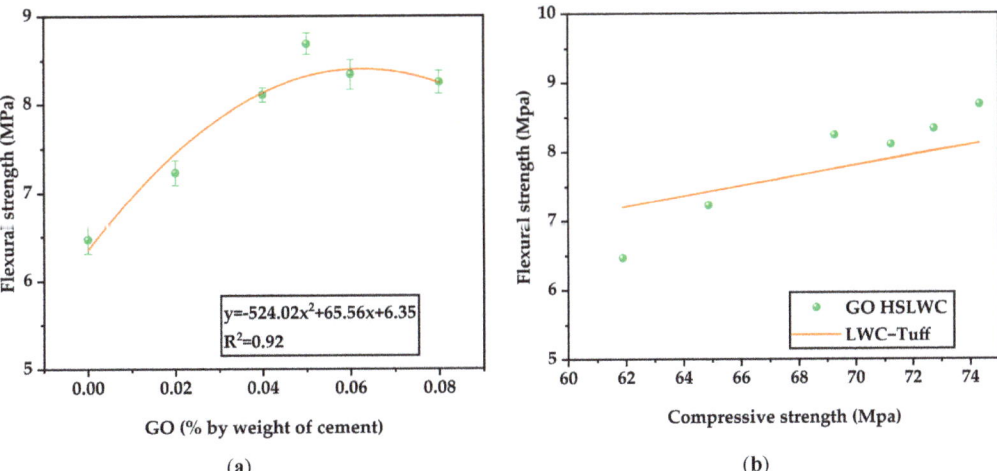

Figure 8. Flexural strength results of HSLWC: (**a**) flexural strength with different contents of GO; (**b**) prediction of flexural strength.

Generally, the flexural strength/compressive strength ratio of HSLWC was around 9–10%, which is lower than that of NWC [47]. Meanwhile, the HSLWC with sintered fly ash obtained a flexural strength/compressive strength ratio of 20% due to modification with steel fiber [48]. The flexural strength/compressive strength ratio (F/C in Table 4) of all mixtures in this study was within the range of 5.9–6.7%. Relevant studies showed that GO had significant strengthening reinforcement effects on flexural strength compared with compressive strength in cement-based materials [28]. This indicated that the improvement in flexural strength was weaker than that in compressive strength, which might have been caused by the brittleness of SC. Figure 8b shows a necessary fitting relationship between compressive strength and flexural strength of HSLWC, and it is recommended that the equations from CEB-FIP for LWC be used for prediction with an estimated error of no more than 10%.

3.1.3. Splitting Tensile Strength

Figure 9a presents the splitting tensile strength of HSLWC specimens with various contents of GO at 28 days under standard curing. Similar to flexural strength, the splitting tensile strength of the HSLWC with GO was evidently higher than that of those without GO. The variation trend in splitting tensile strength with the increase in GO content was similar to that of compressive strength and flexural strength. The splitting tensile strengths of GO-0, GO-2, GO-4, GO-5, GO-6, and GO-8 were 4.21, 4.65, 4.92, 5.23, 5.13, and 4.95 MPa, respectively. As expected, the specimens with GO achieved a maximum increase of 24% when the content of GO reached 0.05%.

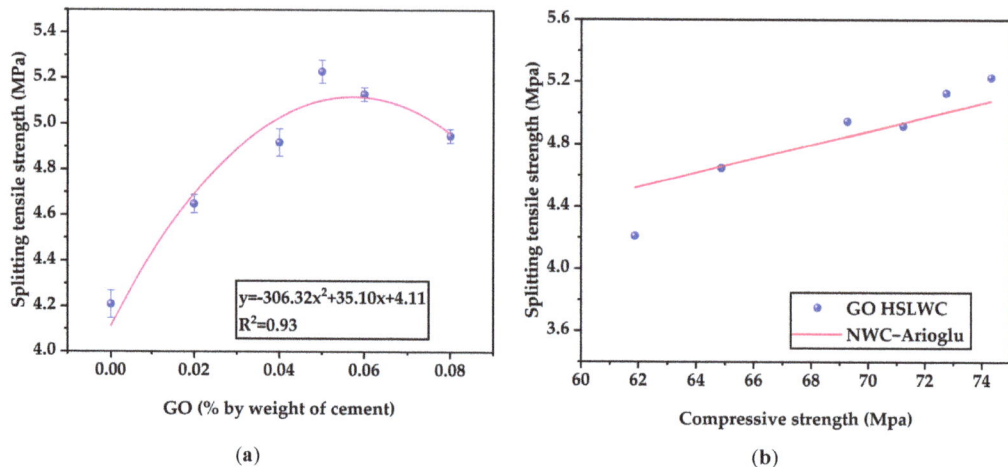

Figure 9. Splitting tensile strength results of HSLWC: (**a**) splitting tensile strength with different contents of GO; (**b**) prediction of splitting tensile strength.

The splitting tensile strength/compressive strength ratio of HSLWC was approximately 6.2–7.1%, while the range for NWC was 8~14% [49]. Lee et al. reported that the ratio of OPBC HSLWC after 28 days of full water curing was 6–7.7% [10]. The splitting tensile strength/compressive strength ratio (S/C in Table 4) of all mixtures in this study was within the acceptable range of 6.8–7.2%. Figure 9b shows a reasonable fitting relationship between compressive strength and splitting tensile strength of HSLWC. Various fitting equations for the prediction of splitting tensile strength are presented. Based on the test results of this study, the equation proposed in the literature obtained a minimum estimation error of less than 8% [50].

3.2. Durability

3.2.1. Chloride Penetration Resistance

Figure 10 shows the chloride-ion migration coefficient results of specimens with different GO incorporation contents at 28 days. Similarly, the specimens incorporating GO obtained lower chloride-ion migration coefficients than those without GO, indicating that incorporating GO into HSLWC significantly promoted the resistance of chloride penetration. With the increase in GO content, the chloride-ion migration coefficients showed a nonlinear trend of first decreasing and then increasing, and the chloride-ion migration coefficient reached a minimum when the GO content was 0.05. This optimal GO content is consistent with the findings of a previous studies by Yu et al. [36]. The chloride-ion migration coefficient values of GO-0, GO-2, GO-4, GO-5, GO-6, and GO-8 were 7.18×10^{-12}, 5.65×10^{-12}, 4.44×10^{-12}, 4.07×10^{-12}, 4.25×10^{-12}, and 4.91×10^{-12} m^2/s, respectively, which suggested that GO can improve the chloride-ion migration coefficient of HSLWC to achieve a maximum reduction of 43%. The inherent key to the resistance of chloride-ion permeability is the pore size of concrete [51]. As mentioned earlier, owing to the filling of nanoscale particles, GO not only refines the structure to form harmless pores but also blocks or cuts off transportation to reduce porosity. On the contrary, GO gradually becomes agglomerated after the addition of more than the optimal content, which, in turn, lead to the deterioration of chloride-ion penetration resistance.

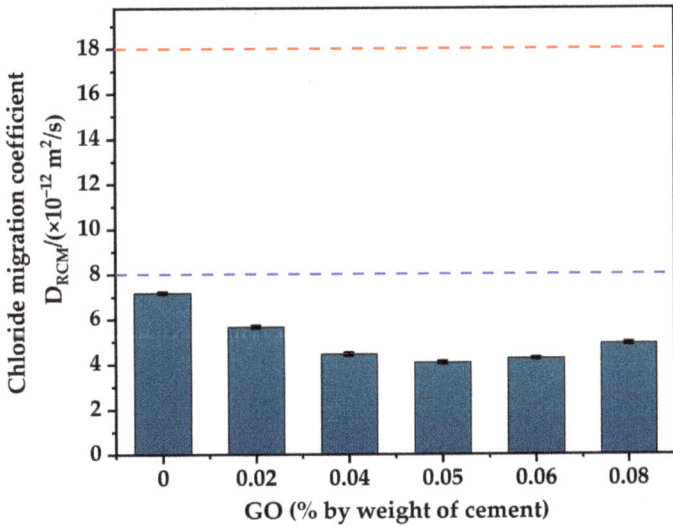

Figure 10. The chloride-ion migration coefficient of HSLWC with different contents of GO.

In addition, based on the resistance level divided by the 28-day chloride-ion migration coefficient, Luping et al. found that concrete with a migration coefficient greater than 18×10^{-12} had poor resistance to marine environments, and concrete with a migration coefficient less than 8×10^{-12} had good resistance to natural environments [52]. GO can provide better corrosion resistance for HSLWC to adapt to marine environments. Hence, it could be inferred that the migration coefficients of all mixtures in the study were within the satisfactory range.

3.2.2. Freezing–Thawing Resistance

Figure 11 presents the freezing and thawing test results of specimens with different GO incorporation contents up to 250 days. As shown in Figure 11a, the mass loss rate of all mixtures gradually increased with the increase in cycles. Meanwhile, the mass losses of the specimens incorporating GO were lower than those of the specimens without GO, which indicated that GO could effectively slow down the mass losses of the during in the freezing and thawing cycles. After 250 cycles, the mass loss rates of GO-0, GO-2, GO-4, GO-5, GO-6, and GO-8 were 2.91%, 2.23%, 1.65%, 1.10%, 1.35%, and 1.92%, respectively. At this time, the surface of the specimens only grew some small holes but did not deteriorate to form cracks or peeling. The mass loss rate curve of GO-5 was higher than the mass loss rate curve of other mixes, indicating that 0.05% GO could reduce the mass loss rate to the greatest extent. As shown in Figure 11b, the relative dynamic elastic modulus had the same variation characteristics as the mass loss rate. After 250 cycles, the relative dynamic elastic moduli of GO-0, GO-2, GO-4, GO-5, GO-6, and GO-8 were 95.4%, 96.2%, 97.3%, 98.4%, 97.9%, and 96.7%, respectively. The dynamic elastic modulus curve of GO-5 was lower than the dynamic elastic modulus curve of other mixes, indicating that 0.05% GO could prevent the attenuation of the dynamic elastic modulus to the greatest extent.

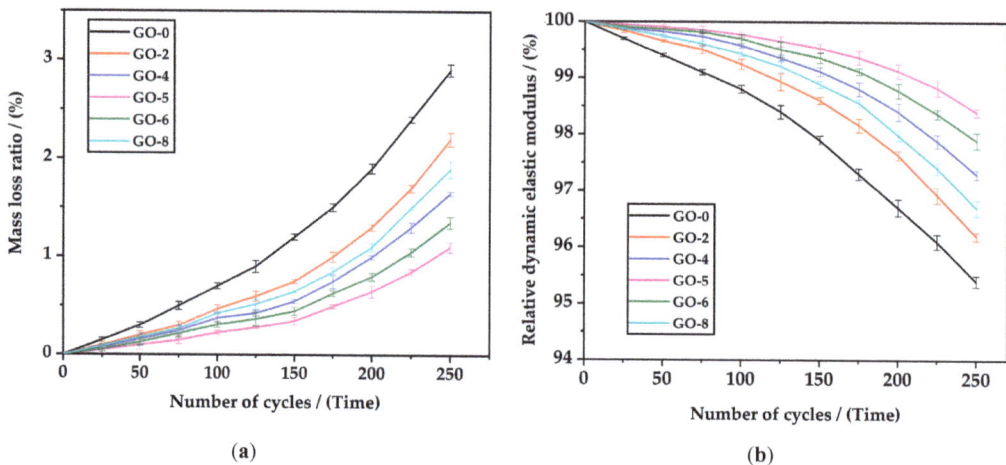

Figure 11. Freezing and thawing results of HSLWC with different GO incorporation contents: (**a**) the mass loss rate; (**b**) the relative dynamic elastic modulus.

The above results showed that 0.05% GO, as the optimal dosage, could help HSLWC obtain the best freezing and thawing resistance in this study. Freezing and thawing damage mainly depends on the swelling of capillary water in micro pores of concrete, whereas the addition of GO can refine pores and reduce porosity, thus impeding the free flow of capillary water. In addition, the mass loss rate and relative elastic modulus of all mixtures were in the range of 1.10–2.91% and 95.4–98.4%, respectively, which met the requirements of GB/T 50082-2009. With the contribution of GO, UHPC can obtain a relatively dynamic modulus in the range of 95.93–98.51% after 300 cycles [36]. It also should be highlighted that all the mixes in this study had excellent freezing and thawing resistance.

3.2.3. Sulfate Attack Resistance

Figure 12 shows the sulfate attack resistance results of HSLWC with different GO incorporation contents up to 150 cycling times in a sulfate solution. As shown in Figure 12a, the mass of all mixtures slightly increased in the first 60 cycles and gradually decreased in the remaining 90 cycles, indicating that the mass loss ratio showed a trend of first negative growth and then positive growth. The specimens without GO had a more significant vibration amplitude in their mass change than those with GO, which was mainly due to the fact that GO prevented deterioration in the wet and dry cycles. After 150 cycles, the mass loss rates of GO-0, GO-2, GO-4, GO-5, GO-6, and GO-8 were 2.95%, 2.21%, 1.65%, 1.54%, 1.87%, and 1.99%, respectively. As shown in Figure 12b, the corrosion resistance coefficient had the same trend as the mass loss rate of first increasing and then decreasing. The corrosion resistance coefficient of the specimens without GO decreased after 60 cycles, while those of the specimens with GO decreased after 90 cycles. After 150 cycles, the corrosion resistance coefficients of GO-0, GO-2, GO-4, GO-5, GO-6, and GO-8 were 86.3%, 89.3%, 93.7%, 97.4%, 94.6%, and 91.3%, respectively. Accordingly, the nonlinear enhancement effect was probably attributable to the uneven dispersion or supersaturation of GO, suggesting that the optimal content for sulfate resistance was 0.05%.

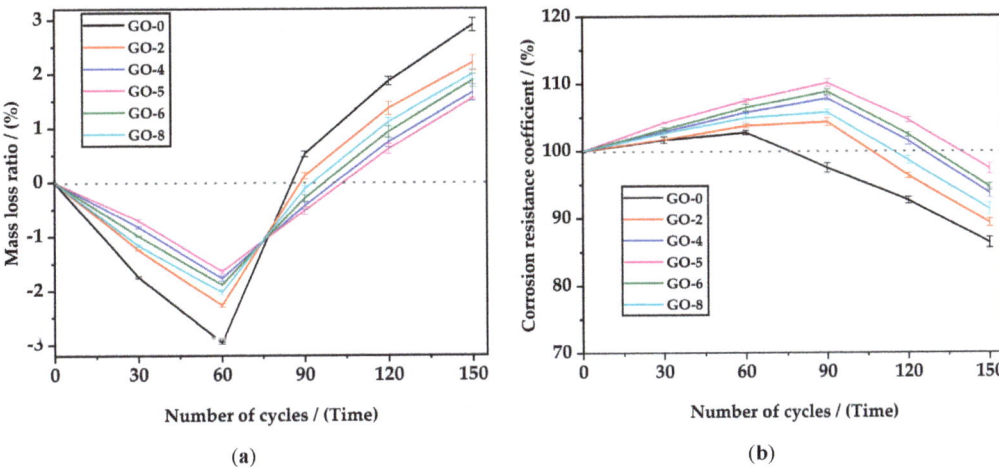

Figure 12. Sulfate attack resistance results of HSLWC with different GO incorporation contents: (**a**) mass loss ratio; (**b**) corrosion resistance coefficient.

Corrosion resistance was reported to be related to ion transport and pore structure [53]. At the early stage of corrosion, ettringite crystals were continuously formed and accumulated, filling capillary pores, thus temporarily increasing the compressive strength and weight. In the later stage of erosion, sulfate can gradually consume and destroy the skeleton of hydration products, thus reducing the compressive strength and weight. Considering the nanofold morphology, GO could block or cut off ion transport, thus mitigating corrosion damage [54]. Hence, all mixtures in this study had excellent corrosion resistance according to the requirements of GB/T 50082-2009.

3.3. Microstructure

Figure 13 shows the SEM images of samples randomly investigated from different mix proportions at 28 days. As shown in Figure 13a, some typical crystals were observed in the specimens without GO. These typical crystals were similar to those produced by hydration reaction in cement mortar, such as layered crystals, rod-like crystals, and sheet-like crystals, which were assembled from the composite formed by AFt, AFm, and CH. This process was also accompanied by the formation of nanoscale pores and microcracks.

The formation and growth of flower-like crystals could be observed from the samples containing GO, as shown in Figure 13b–f. In particular, in contrast with GO-0, many clusters of flower-like crystals formed and grew in an orderly manner in the interfacial transition zones and pores of the GO-2 mixture (Figure 13b). When the GO content increased from 0.04% to 0.06%, the petals of the flower-like crystals grew stronger and gradually matured (Figure 13c–e). When the content of GO reached 0.08%, the shape of the flower crystals were almost unchanged, and the number of flower crystals decreased (Figure 13f). Lv et al. confirmed that GO can participate in the hydration reaction to produce a unique and dense flower-like crystal [24]. Chuah et al. also reported that these flower-shaped crystals were beneficial to improving the mechanical properties and durability of concrete [55].

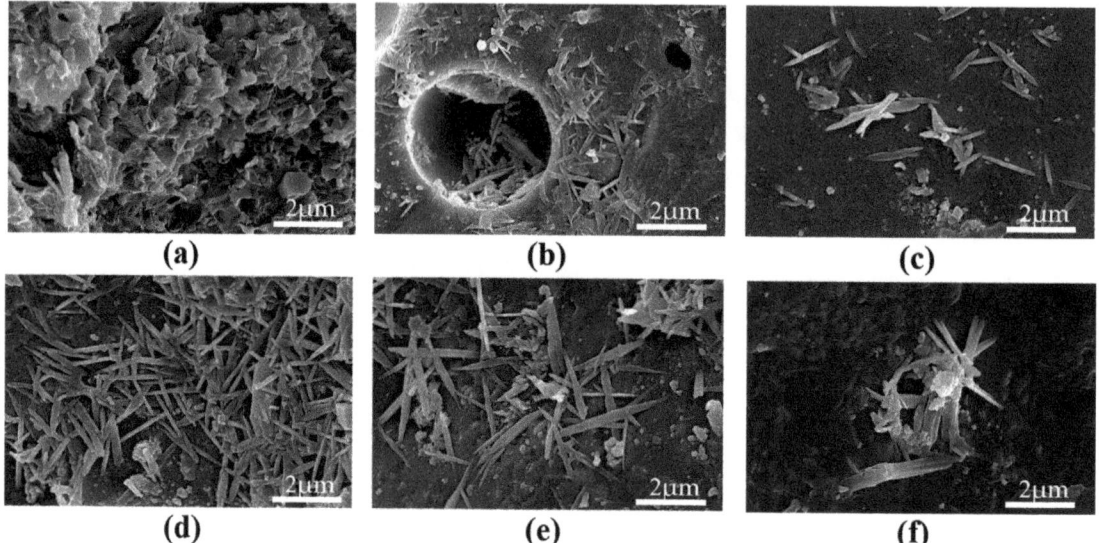

Figure 13. SEM images of different kinds of mix proportions at 28 days: (**a**) GO-0; (**b**) GO-2; (**c**) GO-4; (**d**) GO-5; (**e**) GO-6; (**f**) GO-8.

4. Conclusions

The main objective of this study was to design an initial mixture of HSLWC with SC as an aggregate, and we added six different low contents id GO to compare the enhancement effect of the mechanical properties and durability. In addition, the microstructure of HSLWC with different GO contents was also investigated. The main conclusions of this study are as follows:

- The specimens with different GO contents had an oven-dry density in the range of 1696–1728 kg/m^3 and a compressive strength in the range of 61.88–74.32 MPa, which meet the classification requirements of HSLWC. GO not only adjusted the crystal morphology at an early stage but also maximized the 28-day compressive strength by 20.1%. The specimens with different GO contents had a flexural strength ranging from 6.47 to 8.69 MPa. The addition of GO could increase the flexural strength by 11.7–34.3%. The specimens with different GO contents had a splitting tensile strength ranging from 4.21 to 5.23 MPa. The addition of GO could increase the splitting tensile strength by 10.5–24.2%.
- The chloride-ion migration coefficient of HSLWC with different GO incorporation contents was within the range of 4.07×10^{-12}–7.18×10^{-12} m^2/s, suggesting that the HSLWC in this study could be well applied to marine environments. GO could help the chloride-ion migration coefficient of HSLWC to reach a maximum reduction of 43%. After 250 freezing and thawing cycles, the specimens with different GO contents had a mass loss rate in the range of 1.10–2.91% and a relative dynamic elastic modulus in the range of 95.4–98.4%. After 150 wet and dry cycles, the specimens with different GO contents had a mass loss rate in the range of 1.54–2.95% and a corrosion resistance coefficient in the range of 86.3–97.4%. These results indicated that GO can improve the freeze–thaw resistance and sulfate attack resistance of HSLWC.
- When the content of GO increased from 0 to 0.08%, all the performance indices of HSLWC showed a nonlinear trend. The peak in performance occurred when the GO content was 0.05%. It could be inferred that the optimal GO addition of HSLWC produced from SC was 0.05%. A low content of GO could adjust the crystal morphology to grow flower-like crystals. The number and size of flower-like crystals

- had a nonlinear relationship with the content of GO. This may be another important reason for the observed performance improvement.
- The results indicated that a low content GO could contribute better mechanical properties and durability to HSLWC, thereby extending the service life of buildings and reducing maintenance costs. The addition of different amounts of GO produces different reinforcement effects. GO can be used to achieve the application of SC in high-rise and large-span structures as well as in extreme cold or deep sea areas. Therefore, using GO to strengthen HSLWC made of SC has broad application prospects.
- Oxygen content is an important parameter for the affinity and mechanical properties of GO. Despite the significant mechanical and durability enhancements in this study, controlling the oxygen content of GO to accurately adjust the performance of HSLWC still requires further research to achieve wider practical applications.

Supplementary Materials: The following supporting information can be downloaded at: https://www.mdpi.com/article/10.3390/ma16072756/s1, Formula S1: the chloride ion migration coefficient; Formula S2: the mass loss rate of freezing–thawing resistance; Formula S3: the relative elastic modulus; Formula S4: the mass loss rate of sulfate attack resistance; Formula S5: the corrosion resistance coefficient.

Author Contributions: Conceptualization, X.H. and J.C.L.; methodology, X.H., J.L.N. and J.C.L.; formal analysis, J.L.N.; investigation, X.H., J.C.L. and Z.M.Y.; resources, X.H., Q.H. and Q.L.; writing—original draft preparation, X.H. and Q.H.; writing—review and editing, X.H. and J.C.L.; visualization, X.H., Q.L. and Z.M.Y.; supervision, J.C.L. and J.L.N. All authors have read and agreed to the published version of the manuscript.

Funding: This research was funded by the Doctoral Research Foundation of Xichang University (YBZ202144) and UCSI University Research Excellence & Innovation Grant (REIG) (REIG-FETBE-2022/017).

Institutional Review Board Statement: Not applicable.

Informed Consent Statement: Not applicable.

Data Availability Statement: Data are contained within the article.

Conflicts of Interest: The authors declare no conflict of interest.

References

1. Al-Khaiat, H.; Haque, N. Strength and Durability of Lightweight and Normal Weight Concrete. *J. Mater. Civ. Eng.* **1999**, *11*, 231–235. [CrossRef]
2. Kayali, O. Fly Ash Lightweight Aggregates in High Performance Concrete. *Constr. Build. Mater.* **2008**, *22*, 2393–2399. [CrossRef]
3. Real, S.; Gomes, M.G.; Moret Rodrigues, A.; Bogas, J.A. Contribution of Structural Lightweight Aggregate Concrete to the Reduction of Thermal Bridging Effect in Buildings. *Constr. Build. Mater.* **2016**, *121*, 460–470. [CrossRef]
4. Haque, M.N.; Al-Khaiat, H.; Kayali, O. Strength and Durability of Lightweight Concrete. *Cem. Concr. Compos.* **2004**, *26*, 307–314. [CrossRef]
5. Kockal, N.U.; Ozturan, T. Durability of Lightweight Concretes with Lightweight Fly Ash Aggregates. *Constr. Build. Mater.* **2011**, *25*, 1430–1438. [CrossRef]
6. Aslam, M.; Shafigh, P.; Jumaat, M.Z. Oil-Palm by-Products as Lightweight Aggregate in Concrete Mixture: A Review. *J. Clean. Prod.* **2016**, *126*, 56–73. [CrossRef]
7. Kılıç, A.; Atiş, C.D.; Yaşar, E.; Özcan, F. High-Strength Lightweight Concrete Made with Scoria Aggregate Containing Mineral Admixtures. *Cem. Concr. Res.* **2003**, *33*, 1595–1599. [CrossRef]
8. Rossignolo, J.A.; Agnesini, M.V.C.; Morais, J.A. Properties of High-Performance LWAC for Precast Structures with Brazilian Lightweight Aggregates. *Cem. Concr. Compos.* **2003**, *25*, 77–82. [CrossRef]
9. Thomas, M.; Bremner, T. Performance of Lightweight Aggregate Concrete Containing Slag after 25years in a Harsh Marine Environment. *Cem. Concr. Res.* **2012**, *42*, 358–364. [CrossRef]
10. Chai, L.J.; Shafigh, P.; Bin Mahmud, H. Production of High-Strength Lightweight Concrete Using Waste Lightweight Oil-Palm-Boiler-Clinker and Limestone Powder. *Eur. J. Environ. Civ. Eng.* **2019**, *23*, 325–344. [CrossRef]
11. Mansouri, E.; Manfredi, M.; Hu, J.-W. Environmentally Friendly Concrete Compressive Strength Prediction Using Hybrid Machine Learning. *Sustainability* **2022**, *14*, 12990. [CrossRef]

12. Hossain, K.M.A.; Ahmed, S.; Lachemi, M. Lightweight Concrete Incorporating Pumice Based Blended Cement and Aggregate: Mechanical and Durability Characteristics. *Constr. Build. Mater.* **2011**, *25*, 1186–1195. [CrossRef]
13. Chen, S.-H.; Wang, H.-Y.; Jhou, J.-W. Investigating the Properties of Lightweight Concrete Containing High Contents of Recycled Green Building Materials. *Constr. Build. Mater.* **2013**, *48*, 98–103. [CrossRef]
14. Liu, Q.; Su, L.; Xiao, H.; Xu, W.; Yan, W.M.; Xia, Z. Preparation of Shale Ceramsite Vegetative Porous Concrete and Its Performance as a Planting Medium. *Eur. J. Environ. Civ. Eng.* **2021**, *25*, 2111–2126. [CrossRef]
15. Wu, X.; Wang, S.; Yang, J.; Zhao, J.; Chang, X. Damage Characteristics and Constitutive Model of Lightweight Shale Ceramsite Concrete under Static-Dynamic Loading. *Eng. Fract. Mech.* **2022**, *259*, 108137. [CrossRef]
16. Fan, L.F.; Wang, H.; Zhong, W.L. Development of Lightweight Aggregate Geopolymer Concrete with Shale Ceramsite. *Ceram. Int.* **2023**, *01*, 1–12. [CrossRef]
17. Zhuang, Y.-Z.; Chen, C.-Y.; Ji, T. Effect of Shale Ceramsite Type on the Tensile Creep of Lightweight Aggregate Concrete. *Constr. Build. Mater.* **2013**, *46*, 13–18. [CrossRef]
18. Moreno, D.; Zunino, F.; Paul, Á.; Lopez, M. High Strength Lightweight Concrete (HSLC): Challenges When Moving from the Laboratory to the Field. *Constr. Build. Mater.* **2014**, *56*, 44–52. [CrossRef]
19. Ma, Y.; Zhu, B.; Tan, M. Properties of Ceramic Fiber Reinforced Cement Composites. *Cem. Concr. Res.* **2005**, *35*, 296–300. [CrossRef]
20. Choi, J.; Zi, G.; Hino, S.; Yamaguchi, K.; Kim, S. Influence of Fiber Reinforcement on Strength and Toughness of All-Lightweight Concrete. *Constr. Build. Mater.* **2014**, *69*, 381–389. [CrossRef]
21. Zhang, Y.; Sun, X. Synergistic Effects of Nano-Silica and Fly Ash on the Mechanical Properties and Durability of Internal-Cured Concrete Incorporating Artificial Shale Ceramsite. *J. Build. Eng.* **2023**, *66*, 105905. [CrossRef]
22. Garcés, P.; Fraile, J.; Vilaplana-Ortego, E.; Cazorla-Amorós, D.; Alcocel, E.G.; Andión, L.G. Effect of Carbon Fibres on the Mechanical Properties and Corrosion Levels of Reinforced Portland Cement Mortars. *Cem. Concr. Res.* **2005**, *35*, 324–331. [CrossRef]
23. Babak, F.; Abolfazl, H.; Alimorad, R.; Parviz, G. Preparation and Mechanical Properties of Graphene Oxide: Cement Nanocomposites. *Sci. World J.* **2014**, *2014*, 1–10. [CrossRef]
24. Lv, S.; Ma, Y.; Qiu, C.; Sun, T.; Liu, J.; Zhou, Q. Effect of Graphene Oxide Nanosheets of Microstructure and Mechanical Properties of Cement Composites. *Constr. Build. Mater.* **2013**, *49*, 121–127. [CrossRef]
25. Low, F.W.; Lai, C.W.; Abd Hamid, S.B. Easy Preparation of Ultrathin Reduced Graphene Oxide Sheets at a High Stirring Speed. *Ceram. Int.* **2015**, *41*, 5798–5806. [CrossRef]
26. Shen, Y.; Liu, B.; Lv, J.; Shen, M. Mechanical Properties and Resistance to Acid Corrosion of Polymer Concrete Incorporating Ceramsite, Fly Ash and Glass Fibers. *Materials* **2019**, *12*, 2441. [CrossRef]
27. Yang, H.; Cui, H.; Tang, W.; Li, Z.; Han, N.; Xing, F. A Critical Review on Research Progress of Graphene/Cement Based Composites. *Compos. Part Appl. Sci. Manuf.* **2017**, *102*, 273–296. [CrossRef]
28. Lv, S.; Ma, Y.; Qiu, C.; Zhou, Q. Regulation of GO on Cement Hydration Crystals and Its Toughening Effect. *Mag. Concr. Res.* **2013**, *65*, 1246–1254. [CrossRef]
29. Sui, Y.; Liu, S.; Ou, C.; Liu, Q.; Meng, G. Experimental Investigation for the Influence of Graphene Oxide on Properties of the Cement-Waste Concrete Powder Composite. *Constr. Build. Mater.* **2021**, *276*, 122229. [CrossRef]
30. Shang, Y.; Zhang, D.; Yang, C.; Liu, Y.; Liu, Y. Effect of Graphene Oxide on the Rheological Properties of Cement Pastes. *Constr. Build. Mater.* **2015**, *96*, 20–28. [CrossRef]
31. Gong, K.; Pan, Z.; Korayem, A.H.; Qiu, L.; Li, D.; Collins, F.; Wang, C.M.; Duan, W.H. Reinforcing Effects of Graphene Oxide on Portland Cement Paste. *J. Mater. Civ. Eng.* **2015**, *27*, A4014010. [CrossRef]
32. Lv, S.; Liu, J.; Sun, T.; Ma, Y.; Zhou, Q. Effect of GO Nanosheets on Shapes of Cement Hydration Crystals and Their Formation Process. *Constr. Build. Mater.* **2014**, *64*, 231–239. [CrossRef]
33. Li, W.; Li, X.; Chen, S.J.; Liu, Y.M.; Duan, W.H.; Shah, S.P. Effects of Graphene Oxide on Early-Age Hydration and Electrical Resistivity of Portland Cement Paste. *Constr. Build. Mater.* **2017**, *136*, 506–514. [CrossRef]
34. Chu, H.; Zhang, Y.; Wang, F.; Feng, T.; Wang, L.; Wang, D. Effect of Graphene Oxide on Mechanical Properties and Durability of Ultra-High-Performance Concrete Prepared from Recycled Sand. *Nanomaterials* **2020**, *10*, 1718. [CrossRef]
35. Wu, Y.-Y.; Zhang, J.; Liu, C.; Zheng, Z.; Lambert, P. Effect of Graphene Oxide Nanosheets on Physical Properties of Ultra-High-Performance Concrete with High Volume Supplementary Cementitious Materials. *Materials* **2020**, *13*, 1929. [CrossRef]
36. Yu, L.; Wu, R. Using Graphene Oxide to Improve the Properties of Ultra-High-Performance Concrete with Fine Recycled Aggregate. *Constr. Build. Mater.* **2020**, *259*, 120657. [CrossRef]
37. Zeng, H.; Lai, Y.; Qu, S.; Qin, Y. Graphene Oxide-Enhanced Cementitious Materials under External Sulfate Attack: Implications for Long Structural Life. *ACS Appl. Nano Mater.* **2020**, *3*, 9784–9795. [CrossRef]
38. Suo, Y.; Guo, R.; Xia, H.; Yang, Y.; Zhou, B.; Zhao, Z. A Review of Graphene Oxide/Cement Composites: Performance, Functionality, Mechanisms, and Prospects. *J. Build. Eng.* **2022**, *53*, 104502. [CrossRef]
39. Xu, Y.; Zeng, J.; Chen, W.; Jin, R.; Li, B.; Pan, Z. A Holistic Review of Cement Composites Reinforced with Graphene Oxide. *Constr. Build. Mater.* **2018**, *171*, 291–302. [CrossRef]
40. Alkhouzaam, A.; Qiblawey, H.; Khraisheh, M.; Atieh, M.; Al-Ghouti, M. Synthesis of Graphene Oxides Particle of High Oxidation Degree Using a Modified Hummers Method. *Ceram. Int.* **2020**, *46*, 23997–24007. [CrossRef]

41. Zhang, P.; Sha, D.; Li, Q.; Zhao, S.; Ling, Y. Effect of Nano Silica Particles on Impact Resistance and Durability of Concrete Containing Coal Fly Ash. *Nanomaterials* **2021**, *11*, 1296. [CrossRef] [PubMed]
42. Hong, X.; Lee, J.C.; Qian, B. Mechanical Properties and Microstructure of High-Strength Lightweight Concrete Incorporating Graphene Oxide. *Nanomaterials* **2022**, *12*, 833. [CrossRef]
43. Monteiro, J.J.M.; Mehta, P.K. Ettringite Formation on the Aggregate—Cement Paste Interface. *Cem. Concr. Res.* **1985**, *15*, 378–380. [CrossRef]
44. Shafigh, P.; Chai, L.J.; Mahmud, H.B.; Nomeli, M.A. A Comparison Study of the Fresh and Hardened Properties of Normal Weight and Lightweight Aggregate Concretes. *J. Build. Eng.* **2018**, *15*, 252–260. [CrossRef]
45. Evangelista, A.C.J.; Tam, V.W.Y. Properties of High-Strength Lightweight Concrete Using Manufactured Aggregate. *Proc. Inst. Civ. Eng. Constr. Mater.* **2020**, *173*, 157–169. [CrossRef]
46. Aitcin, P.C.; Haddad, G.; Morin, R. Controlling plastic and autogenous shrinkage in high-performance concrete. *ACI Specif. Publ.* **2004**, *220*, 69–83.
47. Omar, W.; Mohamed, R.N. The Performance of Pretensioned Prestressed Concrete Beams Made with Lightweight Concrete. *J. Civ. Eng.* **2002**, *14*, 60–70.
48. Domagała, L. Modification of Properties of Structural Lightweight Concrete with Steel Fibers. *J. Civ. Eng. Manag.* **2011**, *17*, 36–44. [CrossRef]
49. Shafigh, P.; Johnson Alengaram, U.; Mahmud, H.B.; Jumaat, M.Z. Engineering Properties of Oil Palm Shell Lightweight Concrete Containing Fly Ash. *Mater. Des.* **2013**, *49*, 613–621. [CrossRef]
50. Arıoglu, N.; Girgin, Z.C.; Arıoglu, E. Evaluation of Ratio between Splitting Tensile Strength and Compressive Strength for Concretes up to 120 MPa and Its Application in Strength Criterion. *ACI Mater. J.* **2006**, *103*, 18–24. [CrossRef]
51. Burgos, D.M.; Guzmán, Á.; Torres, N.; Delvasto, S. Chloride Ion Resistance of Self-Compacting Concretes Incorporating Volcanic Materials. *Constr. Build. Mater.* **2017**, *156*, 565–573. [CrossRef]
52. Luping, T.; Nilsson, L.O. Chloride diffusivity in high strength concrete at different ages. *Nordic Concr. Res.* **1992**, *11*, 162–171.
53. Meng, W.; Khayat, K.H. Effect of Graphite Nanoplatelets and Carbon Nanofibers on Rheology, Hydration, Shrinkage, Mechanical Properties, and Microstructure of UHPC. *Cem. Concr. Res.* **2018**, *105*, 64–71. [CrossRef]
54. Wang, Y.; Yang, J.; Ouyang, D. Effect of Graphene Oxide on Mechanical Properties of Cement Mortar and Its Strengthening Mechanism. *Materials* **2019**, *12*, 3753. [CrossRef] [PubMed]
55. Chuah, S.; Pan, Z.; Sanjayan, J.G.; Wang, C.M.; Duan, W.H. Nano Reinforced Cement and Concrete Composites and New Perspective from Graphene Oxide. *Constr. Build. Mater.* **2014**, *73*, 113–124. [CrossRef]

Disclaimer/Publisher's Note: The statements, opinions and data contained in all publications are solely those of the individual author(s) and contributor(s) and not of MDPI and/or the editor(s). MDPI and/or the editor(s) disclaim responsibility for any injury to people or property resulting from any ideas, methods, instructions or products referred to in the content.

Article

Effect of Defects in Graphene/Cu Composites on the Density of States

Song Mi Kim [1], Woo Rim Park [2], Jun Seok Park [1], Sang Min Song [3] and Oh Heon Kwon [4,*]

[1] Department of Safety Engineering, Graduate School, Pukyong National University, Busan 48513, Republic of Korea
[2] Korea Industrial Safety Association, Seoul 08289, Republic of Korea
[3] Samsung Electronics Co., Ltd., Suwon 16677, Republic of Korea
[4] Department of Safety Engineering, Pukyong National University, Busan 48513, Republic of Korea
* Correspondence: kwon@pknu.ac.kr; Tel.: +82-51-629-6469

Abstract: The process of handling and bonding copper (Cu) and graphene inevitably creates defects. To use graphene/Cu composites as electronic devices with new physical properties, it is essential to evaluate the effect of such defects. Since graphene is an ultrathin anisotropic material having a hexagonal structure, an evaluation of graphene/Cu composites containing defects was conducted taking into account the inherent structural characteristics. The purpose of this study is to evaluate defects that may occur in the manufacturing process and to present a usable basic method for the stable design research and development of copper/graphene composites essential for commercialization of copper/graphene composites. In the future, when performing analytical calculations on various copper/graphene composites and defect shapes in addition to the defect conditions presented in this paper, it is considered that it can be used as a useful method considering defects that occur during application to products of desired thickness and size. Herein, density functional theory was used to evaluate the behavior of graphene/Cu composites containing defects. The density of states (DOS) values were also calculated. The analysis was implemented using three kinds of models comprising defect-free graphene and two- and four-layered graphene/Cu composites containing defects. DOS and Fermi energy levels were used to gage the effect of defects on electrical properties.

Keywords: defect; DFT (density functional theory); DOS (density of states); Fermi level; graphene/Cu composites

1. Introduction

Graphene is a two-dimensional nanomaterial with electron mobility of 15,000 cm$^2 \cdot$V$^{-1} \cdot$s^{-1} at room temperature, resistivity of 10^{-6} $\Omega \cdot$cm, carbon–carbon bond length of 0.142 nm, and interlayer separation of 0.33 nm [1–3]. Using graphene in a composite may provide nanodevices with scalability. Many studies have examined graphene composited with copper (Cu), which has excellent electrical conductivity [4]. For example, Dong et al. [5] studied the electrical conductivity and strength of graphene/Cu nanofilms, Wu et al. [6] analyzed the electrochemical properties of graphene/metal composites for use as energy storage devices, and Jiang et al. [7] compared graphene transistors, silicon transistors, and graphene varistors.

However, defects due to the handling of ultrathin graphene inevitably occur when graphene is composited with Cu. Defects that occur during graphene manufacture make it difficult to identify physical mechanisms related to electrical properties. Therefore, to use graphene/Cu composites in a device, it is necessary to evaluate the effect of defects in the graphene/Cu structure while considering the number of laminated layers. Boukhvalov et al. [8] studied the stability of the graphene/Cu surface using density functional theory (DFT) modeling. This theory enables the calculation of multibody systems by viewing electrons as a single set of electron densities. Garcia-Rodrigez et al. [9] conducted a DFT study of Cu nanoparticles adsorbed on defective graphene. Aslanidou et al. [10] generated fewer than ten layers of graphene bonded to electronic devices via chemical vapor deposition (CVD), and Sherrell et al. [11] studied the electrical stimulation of graphene produced by CVD. Gallegos et al. [12] calculate using DFT to state magnetic and electronic changes of Mn in ZnO and Maleki-Ghaleh [13] et al. investigated the characterization and optical properties of nanoparticles and electronic structure by DFT.

To evaluate the effect of defects on the electrical behavior of graphene/Cu composites at the atomic and molecular levels, this study analyzed the electrical properties of the composites and physical mechanisms of defects. The density of states (DOS) relates to electron occupation at a specific energy level, and DOS value indicates the electrical behavior caused by the movement of free electrons (by knowing the energy level at which electrons or holes may exist). The aim of this study is to figure out the effect of defects present in graphene/Cu composites using DFT analysis method. The tool we used is Quantum Espresso which is an open-source software for nanoscale electronic structure calculations and material modeling. The results of this study provide a basis to evaluate the effect of defects on the electrical behavior of graphene/Cu composites.

2. Methodology and Modeling

2.1. Density Functional Theory

Density functional theory [14–20] is the most commonly used method in electronic structure research because it can handle electronic structures in multibody systems, such as atoms, molecules, and condensates, in the form of a single particle. It facilitates the solution of multibody systems by considering electrons as a set of electron densities, rather than considering their position as a wave function of each electron. In DFT analysis, the reliability and accuracy of the calculation can be adjusted using k-points. We used the Monkhorst–Pack approach [21], which is widely applied to converge calculations by reducing k-point error.

Density of states relates to the number of electrons that can enter a given energy level. The number of electrons that can be filled with a given energy level interval is very important for electronic and light-emitting devices whose performances are related to the flow of electrons. The Kohn–Sham equation [22,23] was used in this study as the governing equation to interpret the DFT, and is described by Equation (1) as follows:

$$(-\frac{1}{2}\nabla^2 + V_{eff}[n])\psi_i(r) = \varepsilon_i \psi_i(r) \tag{1}$$

The Kohn–Sham equation has physical meaning, i.e., the total energy of the ground electronic state of the material system is determined only by the electron density of the ground state, and can be obtained directly by expressing the electron density in the form of a function with arbitrary spatial coordinates r for the electron's bottom state.

The Kohn–Sham equation is the same as the single-particle Schrödinger equation for effective potential $V_{eff}[n]$, which is the orbital energy of the Kohn–Sham orbit ε_i; in turn, this is the density of the system of particle n in the Kohn–Sham state. To solve the Kohn–Sham equation, $V_{eff}[n]$ must be known, and can be calculated using Equation (2) as follows:

$$V_{eff}[n] = V_{ext}[n] + \int \frac{n(r')}{|r-r'|} dr' = v_{xc} \tag{2}$$

$$n(r) = \sum_{i=1}^{N} |\psi_i(r)|^2 \tag{3}$$

$V_{ext}[n]$ can be calculated using electrons close to the atomic nucleus and the potentials produced by the atomic nucleus using approximate potentials. Spatial coordinates near r are represented by r' and v_{xc} corresponds to the exchange-correlation potential. To obtain v_{xc}, we need to know the electron density $n(r)$, which is calculated according to Equation (3).

Obtaining the electron density by solving the Kohn–Sham equation makes it possible to reduce the experimental trial and error that may occur in the material design stage, as well as predict the density of the bottom state of the electron.

2.2. DFT Analysis and Modeling

DOS analysis was performed using Quantum Espresso [24] which can evaluate the nanoscale electronic structure and material modeling. Moreover, after visualizing the data obtained from the assumed modeling analysis using the material simulation platform Materials Square [25], the effect of defects in the graphene/Cu composite was analyzed. We set the k-points grid to $6 \times 6 \times 1$. K-point grid is a calculation evaluation condition to obtain more accurate analysis results. Convergence can occur along the k-point grid. This is because it is necessary to evaluate the convergence of the integral for material properties in Brillouin zone of Monkhorst–Pack approach [21]. The higher the number of k-points used, the more accurate the analysis result, but also increases the cost. Therefore, setting an appropriate k-point is important in DOS analysis.

We use a hybrid functional method, the Perdew–Burke–Ernzerhof (PBE) series for the graphene and Cu potential conditions (Table 1) [26]. PBE increases the computational calculation time and requires a lot of memory in CPU to calculate the exchange energy, but it is a method that can improve accuracy and resolve band gap underestimation. PBE is used to obtain analysis results with adequate accuracy and efficiency. PBE also focuses on evaluating the trend of graphene/Cu composites. The cell parameters that we use are also shown in Table 1.

Table 1. Types of pseudo potential and cell parameter used with Quantum Espresso software.

Element	Pseudo Potential Types	Cell Parameter (a, b, c) (Å)	Cell Parameter (α, β, γ) (°)
Cu	C.pbe-n-kjpaw_psl.1.0.0.UPF	14.77, 14.77, 24.5	90°, 90°, 119.98°
Graphene	Cu_pbe_v1.2.uspp.F.UPF	14.77, 14.77, 20	

The analysis model was constructed for a total of eight cases as a Supplementary Information for the readers. It was based on models of individual, separate layers of graphene and Cu. The modeling of graphene/Cu composites was divided into five cases, which were further subdivided into two- and four-layer designs with and without defects (Table 2). Model 1 corresponded to single-layer graphene, and model 2 to single-layer Cu. Model 3 corresponded to the two-layered graphene/Cu composite, and model 4 to the four-layered graphene/Cu composite; this subdivision was used to evaluate the effect of the number of layers. Model 5 inserted defects into the graphene layer in the two-layered graphene/Cu composite, model 6 inserted defects only into the external graphene layer (first layer) in the four-layered graphene/Cu composite, model 7 inserted defects only into the graphene layer in the middle (third) layer, and model 8 was used to evaluate the effect of defects in the external (first) and middle (third) graphene layers.

Table 2. Density functional theory analysis models.

Model	Type	Thickness (Å)	The Number of Atom	Interlayer Distance (Å)	Lattice Parameter
1	Single layer of graphene	20	Carbon, C 72		
2	Single layer of Cu	24.5	Cu 25		
3	Two-layered Cu/graphene composite	46.56	C 72, Cu 25		
4	Four-layered Cu/graphene composite	95.18	C 144, Cu 50		
5	Two-layered Cu/graphene composite containing graphene defects	46.56	C 66, Cu 25	2.06	27.9
6	Four-layered Cu/graphene composite with a first layer containing graphene defects	95.18	C 138, Cu 50		
7	Four-layered Cu/graphene composite with a third layer containing graphene defects				
8	Four-layered Cu/graphene composite with first- and third-layers containing graphene defects		C 132, Cu 50		

The laminate was modeled according to Figure 1. In the two-layered Cu/graphene composite, the Cu layer was arranged on top of the graphene layer, and in the four-layered case, the materials were laminated in the order of graphene, Cu, graphene, and Cu. In the two-layered case, six carbon atoms were removed from the center of the graphene layer (Figure 2).

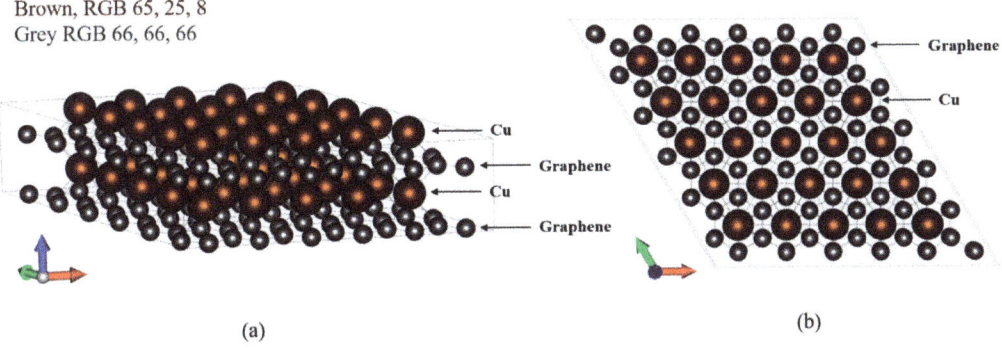

Figure 1. Four-layered graphene/Cu composite model (G/Cu/G/Cu). (**a**) Side view, (**b**) top view.

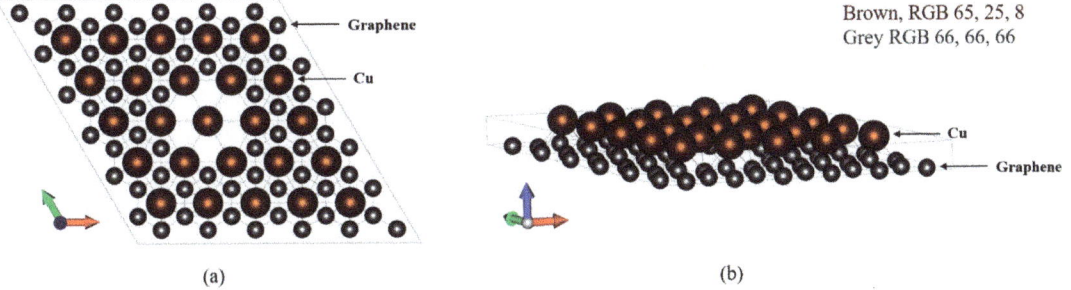

Figure 2. Defect positions of the two-layered graphene/Cu (G/Cu) composite model. (**a**) Side view, (**b**) top view.

In the case of four layers, the graphene layer into which the defects were inserted was changed. We subdivided the case to compare the following designs: four layers with the defects inserted into the external graphene layer (first layer) (Figure 3a), four layers with the defects inserted into the middle graphene layer (third layer) (Figure 3b), and four layers with the defects in both the external graphene layer (first layer), and middle graphene layer (third layer) (Figure 3c).

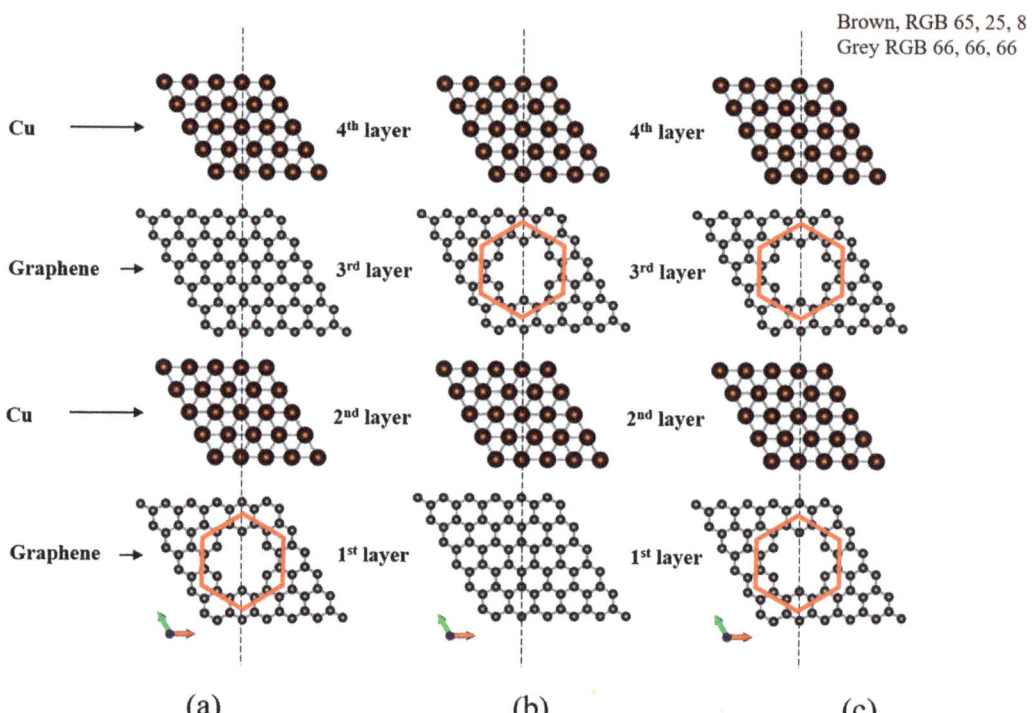

Figure 3. Defect positions of the four-layered Cu/graphene (G/Cu/G/Cu) composite defect model; (**a**) four layers with the defects inserted into the external graphene layer (first layer); (**b**) four layers with the defects inserted into the middle graphene layer (third layer); (**c**) four layers with defects in both the external graphene layer (first layer) and middle graphene layer (third layer).

3. Results and Discussion

3.1. Density Functional Theory Analysis of Individual Graphene and Cu Layers

The entire DOS results of monolayer graphene are presented in Figure 4A, and the PDOS results are presented in Figure 4B which were obtained via DFT analysis of monolayer graphene (model 1). The x-axis corresponds to the energy level function value and the y-axis to DOS value. DOS value of point A (the maximum) ($E - E_f = -6.43$) is 45.36 e/eV^{-1} and DOS value of point B (the minimum) ($E - E_f = 0$) is 0. These values indicate the intrinsic characteristics of graphene. In addition, it is evident that DOS values are relatively evenly distributed, and that the difference between points A and B is <50 e/eV^{-1}. Through Figure 4C,D, it can be seen that point A is due to s orbital, and point B, which is the point where DOS becomes 0 in graphene, is due to p orbital.

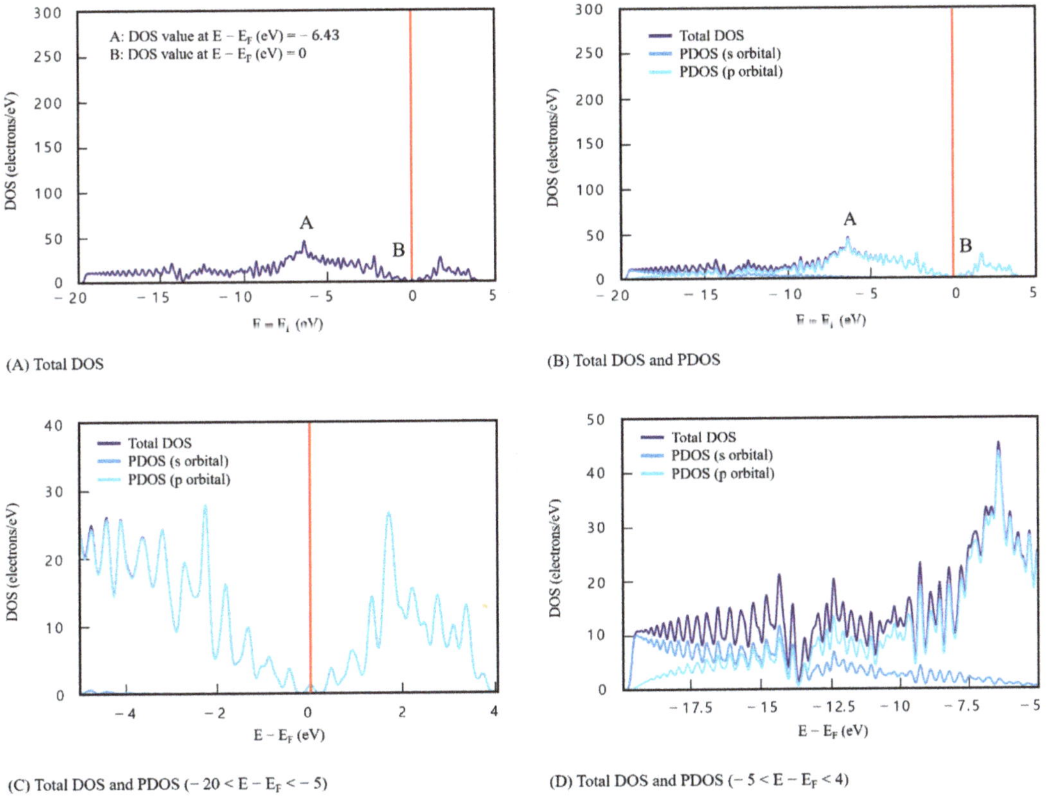

Figure 4. Density of states results for model 1 (single graphene layer, x-axis: $E - E_F$ (eV), y-axis: DOS(electrons/eV), the Fermi level is 0). (**A**) Total DOS. (**B**) Total DOS and PDOS. (**C**) Total DOS and PDOS ($-20 < E - E_F < -5$). (**D**) Total DOS and PDOS ($-5 < E - E_F < 4$).

DOS analysis results for the single layer of Cu (model 2) are presented in Figure 5. A high localized DOS value was observed, which is typical of metals; DOS value at point A (the maximum) ($E - E_f = -1.12$) is 252.02 e/eV^{-1} and that at point B ($E - E_f = 0$) is 9.72 e/eV^{-1}, which indicates that electrons can exist, unlike the graphene results. Moreover, the difference between points A and B is significantly larger (242.3 e/eV^{-1}) compared with graphene. In addition, in model 2, DOS value was negligible before point C ($E - E_f = -5.56$).

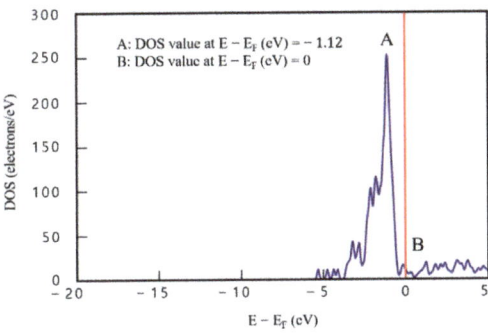

Figure 5. Density of states results for model 2 (single Cu layer).

3.2. Density Functional Theory Analysis of Graphene/Cu Composites

Figure 6 presents DOS analysis results of the two-layered graphene/Cu composite (model 3) consisting of a single layer each of Cu and graphene. In Figure 6, DOS value of point A (the maximum) ($E - E_f = -3.55$) is 182.32 e/eV^{-1} and that of point B ($E - E_f = 0$) is 34.44 e/eV^{-1}, which indicates that electrons could exist at point B, unlike in graphene. The difference between points A and B is 147.88 e/eV^{-1}. In addition, DOS values are evenly distributed over the entire section ($-20 < E - E_f < 5$), confirming the effect of graphene. The difference between point A, where the highest DOS value appears, and point B, where the energy level function value is 0, is approximately 148 e/eV^{-1}, which is about three-times higher than that of graphene. This confirmed the effect of Cu in a section ($-3 < E - E_f < 0$) between points A and B.

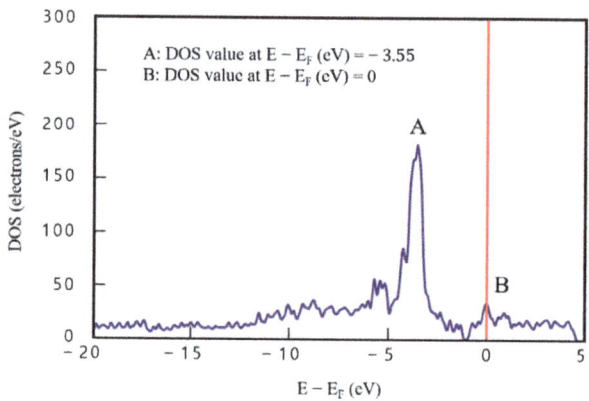

Figure 6. Density of states results for model 3 (two-layered graphene/Cu composite).

Figure 7 shows DOS analysis results of model 4 of the four-layered Cu/graphene composites. DOS value of 209.12 e/eV^{-1} at point A ($E - E_f = -3.55$), which is the same point as for model 3, is 26.8 e/eV^{-1}, higher than that for model 3, and DOS value at point B ($E - E_f = 0$) is 41.46 e/eV^{-1}. Compared with the two-layered case, DOS value of point A is higher by about 14.7%. This finding indicates an improvement in electrical performance, because the number of electrons that can be occupied at the same energy level increased by more than 26 electrons.

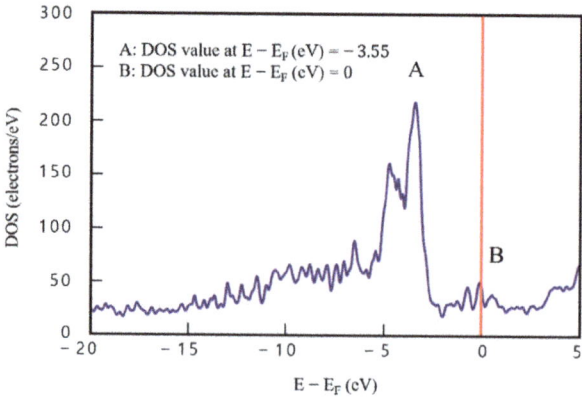

Figure 7. Density of states results for model 4 (four-layered graphene/Cu composite).

In addition, DOS value was quite stable in the sections ($-20 < E - E_f < -5$) and ($3 < E - E_f < 5$), and was more than double compared to that obtained with model 3. These findings indicate that the electrical performance of the graphene/Cu composites improved with increasing graphene/Cu thickness.

3.3. Density Functional Theory Analysis of Graphene/Cu Composites with Graphene Defects

Defects were inserted into the graphene layers of the two- and four-layered graphene/Cu composites to establish the effect of defects on DOS analysis. Figure 8 presents the results for model 5 (Figure 2), in which defects were inserted into graphene in the two-layered graphene/Cu composite. DOS value of point A ($E - E_f = -3.55$) is 140.65 e/eV^{-1} and that of point B ($E - E_f = 0$) is 17 e/eV^{-1}. Compared with the two-layered graphene/Cu composite in which no defect was inserted, point A is lower by about 22.85% (41.67 e/eV^{-1}) and point B is lower by about 50.63% (17.44 e/eV^{-1}). This implies that defects in graphene reduce DOS value and decrease electrical performance.

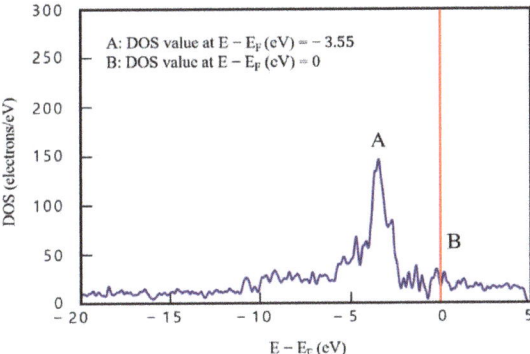

Figure 8. Density of states results for model 5 (two-layered graphene/Cu composite with graphene defects).

Figure 9 presents DOS analysis results for model 6. Point A ($E - E_f = -3.55$) corresponds to 201.87 e/eV^{-1} and point B ($E - E_f = 0$) to 44.2 e/eV^{-1}. Compared with the four-layered graphene/Cu composite without defects, point A is lower by 7.25 e/eV^{-1} (about 3.47%) and point B is higher by 2.74 e/eV^{-1} (about 6.61%). Although defects in the external graphene layer (first layer) in the four-layered graphene/Cu composites led to a slight decrease in DOS value of point A, defects are nevertheless considered to play a doping-like role and increase electrical performance at point B.

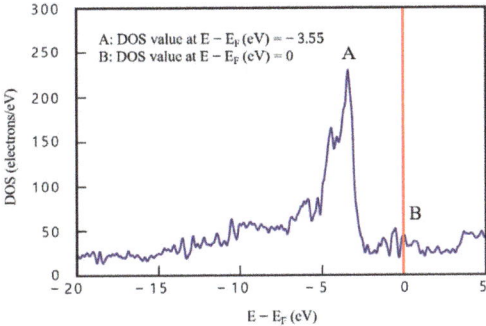

Figure 9. Density of states results for model 6 (four-layered graphene/Cu composite with graphene defects in the first layer).

To examine the effect of these defects in detail, DOS analysis of model 7 was performed by fixing the number of layers and changing only the defect location. Accordingly, the defects were inserted into the middle graphene layer (third layer) instead of the external graphene layer (first layer). The results are presented in Figure 10.

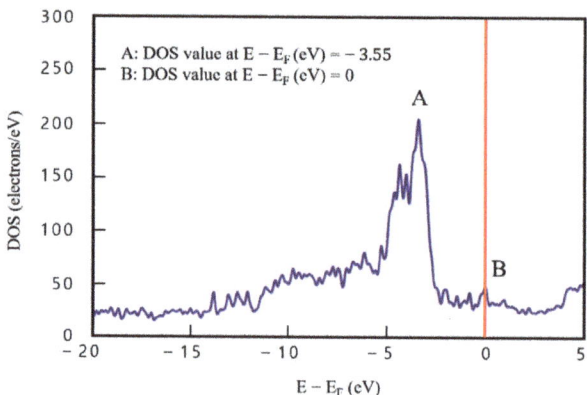

Figure 10. Density of states results for model 7 (four-layered graphene/Cu composite with graphene defects in the third layer).

DOS analysis results of model 7 are presented in Figure 10. Point A ($E - E_f = -3.55$) corresponds to 179.94 e/eV^{-1} and point B ($E - E_f = 0$) to 47.77 e/eV^{-1}. Compared with Figure 7, which presents DOS findings of the defect-free four-layered Cu/graphene composite, Figure 10 shows that point A is lower by 29.18 e/eV^{-1} (about 13.95%), and point B by 6.31 e/eV^{-1} (about 15.22%). In the four-layered graphene/Cu composites, defects in the middle graphene layer (third layer) led to a slight decrease in DOS value of point A, similar to the effect of defects in the external graphene layer (first layer). However, DOS of point B indicates that defects play a doping-like role, facilitating electron movement and thereby improving electrical performance.

The behavior of points A and B in DOS analyses was examined in detail using model 8 (Figure 3c). In this model, defects were inserted into both the external graphene layer (first layer) and middle graphene layer (third layer) of the four-layered graphene/Cu composite. DOS results are presented in Figure 11.

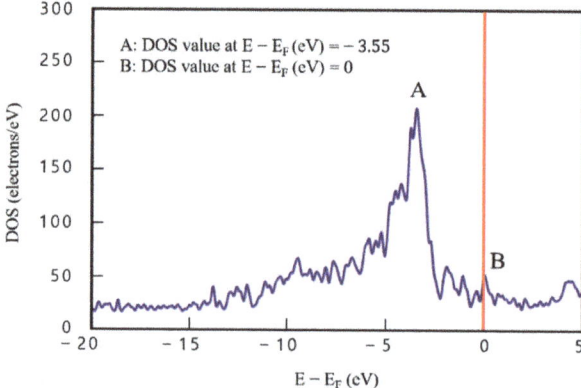

Figure 11. Density of states results for model 8 (four-layered graphene/Cu composite with graphene defects in the first and third layers).

DOS analysis results of model 8 are presented in Figure 11. Point A ($E - E_f = -3.55$) corresponds to 208 e/eV^{-1} and point B ($E - E_f = 0$) to 52 e/eV^{-1}. Figures 9–11 display similar trends. In Figure 11, point A is lower by about 0.05%, and point B is higher by about 25.42%, compared with the defect-free four-layered graphene/Cu composite. Defects in both the external graphene layer (first layer) and middle graphene layer (third layer) in the four-layered graphene/Cu composite led to a slight decrease in DOS value of the maximum (point A), but these defects are considered to improve electrical performance. Overall, the results indicate that defects have a significant positive effect on the electrical performance of graphene/Cu composites at point B of DOS, where the energy level is 0, and that DOS trends higher.

3.4. Effects of Defects in Charge Density of Graphene/Cu Composites

Charge density of model 8 with graphene defects in the first and third layers compared to that of model 5, four-layered graphene/Cu composite, is shown in Figure 12.

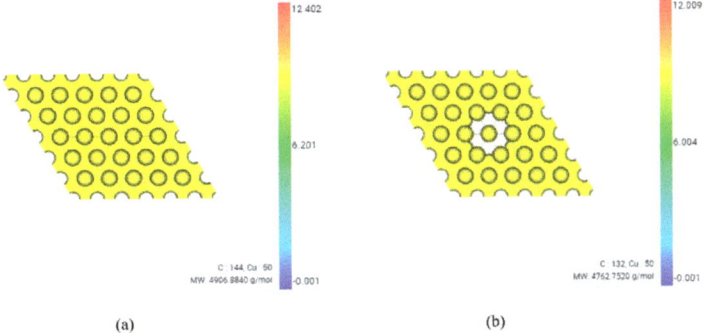

Figure 12. Charge density. (a) model 5 (four-layered graphene/Cu) (b) model 8 (four-layered graphene/Cu composite with graphene defects in the first and third layers).

Molecular weight(MW) of model 5 is 4906.88 g/mol. MW of model 8 is 4762.75 g/mol. In model 5, the range of ρ(max) is 12.402 e/bohr^3 and ρ(min) is −0.00053 e/bohr^3. In model 8, the range of ρ(max) is 12.009 e/bohr^3 and ρ(min) is −0.00056 e/bohr^3. Therefore, it can be seen that the charge density decreases by about 3.167% based on ρ(max) due to defects. It is judged that this is because the existence of defects prevented the bonding force of electrons.

4. Conclusions

DOS analyses of graphene/Cu composites containing defects were conducted to estimate the effect of defects on electrical performance. The following conclusions were drawn:

- DOS behavior of the two-layered graphene/Cu composite (model 3) resembled that of a single graphene layer, i.e., a stable DOS value was observed across the entire section of $-20 < E - E_f < 5$ for the composite and across $-3 < E - E_f < 0$ for Cu. This indicated that DOS of the two-layered graphene/Cu composite was affected by both graphene and Cu.
- Compared with the two-layered case, DOS value of the four-layered graphene/Cu composite was higher by ≥14.69% at point A, having the highest DOS value. Notably, DOS values over the sections $-20 < E - E_f < -5$ and $3 < E - E_f < 5$ were double, which would be expected to improve electrical performance.
- In the two-layered graphene/Cu composite with graphene defects, point A was lower by about 22.85%, and point B by 50.63%, compared with the defect-free two-layered graphene/Cu composite.
- In the four-layered graphene/Cu composite containing defects in the external graphene layer (first layer), DOS value at point A was about 3.47% lower than that of the defect-

free four-layered graphene/Cu composite, while DOS value of point B was higher by about 6.61%. In the four-layered graphene/Cu composite containing defects in the middle graphene layer (third layer), point A was lower by about 13.95% and B was higher by about 15.22%. Therefore, it is likely that the presence of graphene defects in four-layered graphene/Cu composites decreases the number of electrons at point A, and plays a doping-like role to increase electrical performance at point B. In addition, the external graphene defects performed better than the middle graphene defects at point A, but not at point B.

- In the four-layered graphene/Cu composites containing defects both in the external graphene layer (first layer) and middle graphene layer (third layer), DOS value of point A was only about 0.05% lower than that for the defect-free four-layered graphene/Cu composite, while the value of point B was about 25.42% higher. This means that the graphene defects decreased the number of electrons at point A and played a doping-like role, facilitating movement of electrons at point B and thereby improving electrical performance.

This study presents basic data concerning the electrical behavior of graphene/Cu composites containing defects. Understanding the influence of such defects is essential for the commercialization of graphene, and to control the physical properties of graphene/Cu composites.

Supplementary Materials: The following supporting information can be downloaded at: https://www.mdpi.com/article/10.3390/ma16030962/s1, File S1: The coordinates of Model 1 to 8.

Author Contributions: O.H.K.: supervision, conceptualization, visualization, project administration, and writing—review and editing. S.M.K.: methodology, formal analysis, investigation, data curation, visualization, and writing—original draft. W.R.P.: investigation and writing—review and editing. J.S.P.: investigation. S.M.S.: project administration, investigation. All authors have read and agreed to the published version of the manuscript.

Funding: This research was supported by the Basic Science Research Program through the National Research Foundation of Korea (NRF) funded by the Ministry of Education under grant number 2020R1F1A106863511.

Institutional Review Board Statement: Not applicable.

Informed Consent Statement: Not applicable.

Data Availability Statement: The data presented in this study are available on request from the corresponding author.

Conflicts of Interest: The authors declare no conflict of interest.

References

1. Gein, A.K.; Novoselov, K.S. The Rise of Graphene. *Nat. Mater.* **2007**, *6*, 183–191.
2. Verma, D.; Gope, P.C.; Shandilya, A.; Gupta, A. Mechanical-Thermal-Electrical and Morphological properties of Graphene Reinforced Polymer Composites: A Review. *Trans. Indian Inst. Met.* **2014**, *67*, 803–816. [CrossRef]
3. Castro Neto, A.H.; Guinea, F.; Peres, N.M.R.; Novoselov, K.S.; Geim, A.K. The Electronic Properties of Graphene. *Rev. Mod. Phys.* **2009**, *81*, 109–162. [CrossRef]
4. Zhuo, Q.; Mao, Y.; Lu, S.; Cui, B.; Yu, L.; Tang, J.; Sun, J.; Yan, C. Seed-Assisted Synthesis of Graphene Films on Insulation Substrate. *Materials* **2019**, *12*, 1376. [CrossRef]
5. Dong, Z.; Peng, Y.; Tan, Z.; Fan, G.; Guo, Q.; Li, Z.; Xiong, D. Simultaneously enhanced electrical conductivity and strength in Cu/graphene/Cu sandwiched nanofilm. *Scr. Mater.* **2020**, *187*, 296–300. [CrossRef]
6. Wu, Z.; Zhou, G.; Yin, L.; Ren, W.; Li, P.; Cheng, H. Graphene/metal oxide composite electrode materials for energy storage. *Nano Energy* **2012**, *1*, 107–131. [CrossRef]
7. Jhang, S.H.; Jeong, N.B.; Park, D.H.; Chung, H.J. Graphene Electronic Devices: Transistor vs. Barristor. *New Phys. Seae Mulli* **2016**, *66*, 1201–1209. [CrossRef]
8. Boukhvalov, D.W.; Zhidkovc, I.S.; Kukharenko, A.I.; Slesarev, A.I.; Zatsepin, A.F.; Cholakh, S.O.; Kurmaev, E.Z. Stability of boron-doped graphene/copper interface: DFT, XPS and OSEE studies. *Appl. Surf. Sci.* **2018**, *441*, 978–983. [CrossRef]

9. García-Rodríguez, D.E.; Mendoza-Huizar, L.H.; Díaz, C. A DFT study of Cu nanoparticles adsorbed on defective graphene. *Appl. Surf. Sci.* **2017**, *412*, 146–151. [CrossRef]
10. Aslanidou, S.; Garcia-Garcia, A.; Godignon, P.; Rius, G. Electronic interface and charge carrier density in epitaxial graphene on silicon carbide. A review on metal–graphene contacts and electrical gating. *APL Mater.* **2020**, *8*, 100702. [CrossRef]
11. Sherrell, P.C.; Thompson, B.C.; Wassei, J.K.; Gelmi, A.A.; Higgins, M.J.; Kaner, R.B.; Wallace, G.G. Maintaining Cytocompatibility of Biopolymers Through a Graphene Layer for Electrical Stimulation of Nerve Cells. *Adv. Funct. Mater.* **2013**, *23*, 769–776. [CrossRef]
12. Gallegos, M.V.; Luna, C.R.; Peluso, M.A.; Damonte, L.C.; Sambeth, J.E.; Jasen, P.V. Effect of Mn in ZnO using DFT calculations: Magnetic and electronic changes. *J. Alloys Compd.* **2019**, *795*, 254–260. [CrossRef]
13. Maleki-Ghaleh, H.; Sharkeri, M.S.; Dargahi, Z.; Kavanlouei, M.; Garabagh, H.; Moradpur-Tari, E.; Yourdkhani, A.; Fallah, A.; Zarrabi, A.; Koc, B.; et al. Characterization and optical properties of mechanochemically synthesized molybdenum-doped rutile nanoparticles and their electronic structure studies by density functional theory. *Mater. Today Chem.* **2022**, *24*, 100820. [CrossRef]
14. Thomas, L.H. The calculation of atomic fields. Mathematical Proceedings of the Cambridge Philosophical Society. *Math. Proc. Camb. Philos. Soc.* **1926**, *23*, 542–548. [CrossRef]
15. Fermi, E. Eine statistische Methode zur Bestimmung einiger Eigenschaften des Atoms und ihre Anwendung auf die Theorie des periodischen Systems der Elemente. *Z. Phys.* **1928**, *48*, 73. [CrossRef]
16. Dirac, P.A.M. Note on Exchange Phenomena in the Thomas Atom. *Math. Proc. Camb. Philos. Soc.* **1930**, *26*, 376–385. [CrossRef]
17. Hohenberg, P.; Kohn, W. Inhomogeneous Electron Gas. *Phys. Rev. J. Arch.* **1964**, *136*, B864. [CrossRef]
18. Vanin, M.; Mortensen, J.J.; Kelkkanen, A.K.; Garcia-Lastra, J.M.; Thygesen, K.S.; Jacobsen, K.W. Graphene on metals: A van der Waals density functional study. *Phys. Rev. B* **2010**, *81*, 081408. [CrossRef]
19. Kittel, C.; McEuen, P.; McEuen, P. *Introduction to Solid State Physics*; Wiley: New York, NY, USA, 1996.
20. Ibach, H.; Lüth, H. *Solid-State Physics*; Springer: Berlin, Germany, 2003.
21. Monkhorst, H.J.; Pack, J.D. Special points for Brillouin-zone integrations. *Phys. Rev. B* **1976**, *13*, 5188–5192. [CrossRef]
22. Kohn, W.; Sham, L.J. Self-consistent Equations including Exchange and Correlation Effects. *Phys. Rev.* **1965**, *140*, A1133. [CrossRef]
23. Jang, H.M. Density functional theory and its application to ferroelectric/multiferroelectric studies. *Ceramist* **2015**, *18*, 26–43.
24. Giannozzi, P.; Baroni, S.; Bonini, N.; Calandra, M.; Car, R.; Cavazzoni, C.; Ceresoli, D.; Chiarotti, G.L.; Cococcioni, M.; Dabo, I.; et al. QUANTUM ESPRESSO: A modular and open-source software project for quantum simulations of materials. *J. Phys. Condens. Matter* **2009**, *21*, 395502. [CrossRef] [PubMed]
25. Materials Square, Virtual Lab. Available online: https://www.materialssquare.com/ (accessed on 31 December 2022).
26. Perdew, J.P.; Burke, K.; Ernzerhof, M. Perdew, burke, and ernzerhof reply. *Phys. Rev. Lett.* **1998**, *80*, 891. [CrossRef]

Disclaimer/Publisher's Note: The statements, opinions and data contained in all publications are solely those of the individual author(s) and contributor(s) and not of MDPI and/or the editor(s). MDPI and/or the editor(s) disclaim responsibility for any injury to people or property resulting from any ideas, methods, instructions or products referred to in the content.

Article

Rheological Investigation of Welding Waste-Derived Graphene Oxide in Water-Based Drilling Fluids

Rabia Ikram [1,*], Badrul Mohamed Jan [1,*], Waqas Ahmad [2], Akhmal Sidek [3], Mudasar Khan [2] and George Kenanakis [4,*]

1. Department of Chemical Engineering, Faculty of Engineering, University of Malaya, Kuala Lumpur 50603, Malaysia
2. Institute of Chemical Sciences, University of Peshawar, Peshawar 25120, Khyber Pukhtunkhwa, Pakistan
3. Petroleum Engineering Department, School of Chemical and Energy Engineering, Faculty of Engineering, Universiti Teknologi Malaysia, Johor Bahru 81310, Malaysia
4. Institute of Electronic Structure and Laser, Foundation for Research and Technology-Hellas, N. Plastira 100, Vasilika Vouton, GR-700 13 Heraklion, Crete, Greece
* Correspondence: raab@um.edu.my (R.I.); badrules@um.edu.my (B.M.J.); gkenanak@iesl.forth.gr (G.K.)

Abstract: Throughout the world, the construction industry produces significant amounts of by-products and hazardous waste materials. The steel-making industry generates welding waste and dusts that are toxic to the environment and pose many economic challenges. Water-based drilling fluids (WBDF) are able to remove the drill cuttings in a wellbore and maintain the stability of the wellbore to prevent formation damage. To the best of our knowledge, this is the first study that reports the application of welding waste and its derived graphene oxide (GO) as a fluid-loss additive in drilling fluids. In this research, GO was successfully synthesized from welding waste through chemical exfoliation. The examination was confirmed using XRD, FTIR, FESEM and EDX analyses. The synthesized welding waste-derived GO in WBDF is competent in improving rheological properties by increasing plastic viscosity (PV), yield point (YP) and gel strength (GS), while reducing filtrate loss (FL) and mud cake thickness (MCT). This study shows the effect of additives such as welding waste, welding waste-derived GO and commercial GO, and their amount, on the rheological properties of WBDF. Concentrations of these additives were used at 0.01 ppb, 0.1 ppb and 0.5 ppb. Based on the experiment results, raw welding waste and welding waste-derived GO showed better performance compared with commercial GO. Among filtration properties, FL and MCT were reduced by 33.3% and 39.7% with the addition of 0.5 ppb of raw welding-waste additive, while for 0.5 ppb of welding waste-derived GO additive, FL and MCT were reduced by 26.7% and 20.9%, respectively. By recycling industrial welding waste, this research conveys state-of-the-art and low-cost drilling fluids that aid in waste management, and reduce the adverse environmental and commercial ramifications of toxic wastes.

Keywords: industrial waste; nanomaterials; drilling fluids; graphene oxide; rheology; filtrate loss

1. Introduction

The construction and steel industries produce significant amounts of by-products and welding wastes around the planet [1]. Welding waste is produced during the welding process in the form of vitreous material or slag, of red or black color, which is a non-biodegradable, toxic and hazardous waste material [2,3]. These wastes can be recycled and reused as a constructive material by the proper management of welding waste [4]. For instance, welding wastes can be reused by extracting the valuable minerals inside them through physical or chemical mineral-processing techniques [5]. In addition, welding wastes can be utilized in land filling and in the production of cement [6]. On construction sites, welding waste can be added to road surfaces or the pavements of airports as an asphalt mixture additive [7]. Thus, recycling welding wastes not only conserves the mineral resources but also protects the environment [8].

Steel wastes have similar properties to bentonite clay. Bentonite clay is widely used in the petroleum industry. Almost 80% of drilling fluids consist of bentonite clay. Drilling fluids perform many functions in the drilling operation, such as transporting the drill cuttings from the bottom of a wellbore to the surface; suspending cuttings; maintaining the wellbore pressure of the formations to prevent blow out; cooling and lubricating the drill bit and sealing permeable formations [9]. High amounts of aluminum and titanium dioxide (TiO_2) are found in welding waste and can improve the rheological properties of WBDF [10]. Such waste contains similar ingredients to commercial fluid-loss additives, for example, TiO_2. The addition of TiO_2 to drilling fluids increases viscosity, while reducing FL and MCT. Due to its unique properties, TiO_2 enhances thermal conductivity and acts as a viscosifier as compared to conventional drilling fluids [11].

Previous data shows that total steel demand grew to hundreds of millions of metric tons worldwide [12]. In metallurgical processes, different types of slags are generated as by-products or large amounts of residues in metal incineration processes [13]. A noteworthy analysis was performed to investigate solid waste composition and generation rates in six vocational college welding workshops in Malaysia [14]. The data revealed that welding waste was composed of scrap metal, metal dust, welding electrodes and grinding disks which constituted 92.89, 3.64, 3.07 and 0.4 percent of the total welding waste, respectively. The total welding waste generation rates varied from 59.57 to 117.63 kgw^{-1} across the study workshops, with an average of 83.42 kgw^{-1}. Per capita generation rates varied from 0.60 to 1.90 kgw^{-1}, with an average of 1.23 kgw^{-1}. These data showed the potential and environmental effects of welding waste due to the presence of hazardous constituents which were known to contain a variety of metals and metal oxides [14,15].

In recent years, the usage of GO as a drilling mud additive was established. Due to its high surface area, stability, water resistance and strong mud formation properties, it possesses the super-efficient capability of preventing the leakage of drilling mud into the wellbore [16,17]. A variety of studies presented the addition of GO as an emerging additive for WBDF to enhance rheological properties, such as FL and MCT. Nevertheless, none of the previous studies evaluated the effects of welding waste, or its derivatives as additives, in drilling fluids [18,19]. By choosing the optimal concentration of GO, the properties and hydraulics of drilling fluids can be enhanced [20]. GO helps create an ideal drilling mud as it can prevent the invasion of drilling mud into the formation. Thus, GO-based drilling fluid can significantly reduce friction between the borehole and the drill pipe due to its fine film-forming capabilities [21].

Drilling fluid serves as the material to cool the drill bit during a long drilling operation. When there is contact between the drilling fluids and formation, minimal impact on the mechanical properties of the formation itself is essential. [22]. In order to complete a drilling operation successfully, maintaining an open hole is very important. The addition of welding waste, and its derived GO as additives, allows the drilling fluid to serve as a hole cleaner to eliminate the cuttings from the bottom of the well hole and to control subsurface pressure [23]. The usage of bentonite and different expensive additives used in drilling fluid formulations is expected to decrease as these are replaced with sustainable waste-derived additives. Moreover, this will reduce the residual effects of welding and construction waste, thereby protecting the environment [24]. FL is commonly known as the loss of the mud filtrate into the porous permeable formation. The invasion of drilling fluid is due to the presence of a higher hydrostatic pressure in the wellbore than the pressure in the formation. This is a critical issue during a drilling operation as it leads to many problems, such as formation damage and pipe sticking. If mud filtrate invades the formation, production of the hydrocarbon decreases as the permeability, capillary pressure and wettability are affected. Thus, non-toxic, low cost and environmentally safe additives are added to the drilling fluids to reduce fluid loss into the formation [25].

The application of waste materials in the petroleum industry is important to reduce the negative impact on society. Conventional drilling fluids produce thick mud cake and a significant amount of filtrate loss (FL) which eventually damage the reservoir [26]. Furthermore, current industry practice uses commercial GO which is expensive at around MYR 1000 per few grams, which increases the investment cost of drilling operations. The availability of carbonaceous industrial waste also triggers the initiative to turn such industrial waste into value-added products [27]. To the best of our knowledge, this is the first study using welding waste and welding-waste-derived GO in WBDF. The aim of the research is to recognize the usage of welding waste to improve its filtration properties as a fluid loss agent. This work, therefore, endeavors to fulfill the research gap by proposing the recycling of welding waste and its derived GO to investigate the effect of drilling fluids, compared to basic water bentonite suspensions. The performance of rheological and fluid loss properties was conducted by incorporating welding waste and welding waste-derived GO and comparing them to commercial GO prepared at three different levels (low/medium/high 0.01, 0.1 and 0.5 wt%). Firstly, the unwanted industrial welding waste was converted into GO using a modified Hummers method. Characterization of the fabricated GO was then performed using XRD, FTIR, FESEM and EDX analysis. Furthermore, these additives successfully demonstrated the effectiveness of waste additives following examination of their rheological properties under American Petroleum Institute (API) standards; the properties included plastic viscosity (PV), yield point (YP), gel strength (GS), FL and mud cake thickness (MCT). Finally, we also determine the optimized amount of welding waste and welding waste-derived GO additives that produced the lowest FL and thinnest mud cake.

2. Materials and Methods

2.1. Chemicals and Reagents

All the chemicals used were of analytical grade and used as received. Sulfuric Acid (H_2SO_4 98%), Hydrochloric Acid (HCl 37%) and Sodium Nitrate ($NaNO_3$) were purchased from Riedel-de-Haen. Potassium Permanganate ($KMnO_4$) was supplied by Merck, and Potassium Hydroxide (KOH) and Hydrogen peroxide solution 35% (H_2O_2) from Sigma-Aldrich. Deionized water was used throughout the experiments.

The following components of drilling fluids were used: Sodium hydroxide (purity $\geq$ 99 wt%), Potassium chloride (purity $\geq$ 99 wt%) and Carboxymethylcellulose (purity $\geq$ 99 wt%) were acquired from Sigma-Aldrich, St. Louis, MO, USA. Bentonite and barite (purity 91–93 wt%) were provided by M-I SWACO, Malaysia. Distilled water was used to prepare all aqueous solutions with no further purification.

2.2. Synthesis of Welding Waste-Derived GO

A welding-waste sample was collected from a local welding workshop. The waste sample was a gray colored powder material with a high density due to the presence of metallic residues. The welding-waste sample was first carbonized in a muffle furnace, in the absence of oxygen, at 300 °C for 3 h. The carbonized char was ground to a fine powder and screened through a 125 μm mesh size sieve. The carbonized mass was then pyrolyzed under continuous flow of N_2 in a stainless-steel tubular reactor. About 5 g of sample was placed in the reactor tube which was connected to the N_2 supply. The tube was inserted into a tubular furnace and heated at 550 °C for 1 h. The sample was allowed to cool within the reactor and then stored in vials [28].

To remove the metallic, mineral and ash residue, the sample was first treated with KOH solution (in a ratio of 1:1.4 w/w) under vigorous treatment for 4 h in a beaker. The suspension was allowed to settle for a few hours and the clear aqueous layer was removed. The sample was then suspended in distilled water under vigorous stirring; the suspension layer was quickly decanted and the bottom residues were discarded. The suspension was allowed to settle and then filtered to allow the residue to be collected. The residue was excessively washed with 0.1 N HCl solution and distilled water until the pH of washing was neutralized. The pyrolyzed carbon residue was dried in an oven at 60 °C for 5 h and then stored in vials.

GO was prepared from pyrolyzed carbon powder via an improved Hummers' method [16,29,30]. About 5 g of pyrolyzed carbon derived from welding waste was added to a flask containing a solution of concentrated H_2SO_4 (92 mL) and $NaNO_3$ (4 g), placed in an ice bath and stirred for 10 min. About 15 g of $KMnO_4$ was slowly added to the suspension under continuous stirring; the temperature rose due to the exothermic nature of the reaction, and was maintained at less than 30 °C in the ice bath. After this, 2–3 mL of H_2O_2 was dropwise added to the suspension and the temperature rapidly rose but was controlled to about 90 °C. The color of the slurry turned brown; the flask was covered and allowed to stand overnight. About 180 mL of distilled water was added to the mixture and slowly heated to 90–95 °C for 30 min; the brown-colored residue (GO) was collected through filtration, excessively washed with distilled water and then dried in oven at 60 °C for 5 h [31]. The sequences of steps involved in the synthesis of GO are shown in Figure 1 below.

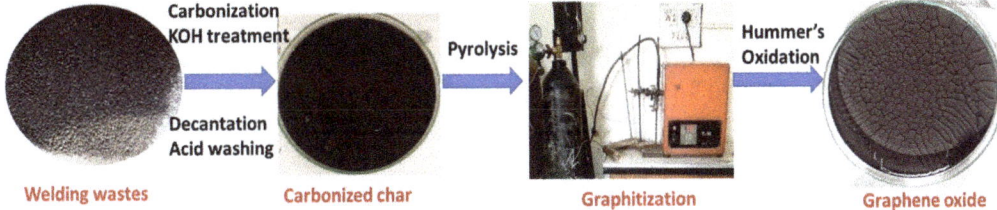

Figure 1. Stepwise synthesis of GO from welding wastes.

2.3. Characterization of GO

The welding waste-derived GO was characterized by FTIR, FESEM, EDX and XRD analysis. The composition and crystallinity of GO was investigated through X-ray diffractometer (XRD; model JDX-9C, JOEL, Tokyo, Japan) using CuKα radiation (1.54178 A° wavelength) and an Ni filter. The surface morphology and elemental composition was studied by FESEM and EDX analysis through Scanning Electron Microscope (Model JEOL-Jsm-5910; Tokyo, Japan). The functional group composition was evaluated by FTIR spectrophotometer (Schimadzu-A60, Tokyo, Japan).

2.4. Drilling Fluid Preparation

Three types of WBDF were prepared in this research; these were basic WBDF with (1) commercial GO, (2) raw welding waste and (3) welding waste-derived GO as the additives. The different additives were added to the base fluid using different concentrations—0.01, 0.1 and 0.5 ppb—as described in previous research work [32]. During the formulation of the base fluid, water was added to the bentonite to create a hydrated slurry, followed by the addition of the remaining components. Table 1 shows the formulation of the WBDF including each material, required mixing time and mixing order for all additives.

Table 1. Formulations of WBDF.

Materials	Basic WBDF	Basic WBDF + Waste Additives	Mixing Time	Mixing Order
Distilled water (mL)	300.0	300.0	-	1
Bentonite (ppb) [$Al_2O_3.4(SiO_2).H_2O$]	25.0	25.0	5	2
Potassium chloride (ppb)	20.0	20.0	2	3
Xanthan gum (ppb)	1.0	1.0	5	4
Polyanionic cellulose (ppb)	3.0	3.0	5	5
Sodium hydroxide (ppb)	0.1	0.1	5	6
Barite (ppb)	70.0	70.0	20	7
Welding waste (ppb)/ Welding waste-derived GO (ppb)/ Commercial GO (ppb)	-	0.01, 0.1, 0.5	5	8

2.5. Evaluation of Rheological and Filtration Properties

A Fann Model 35 Viscometer (Houston, TX, USA) and an Anton Paar rheometer (Germany) were operated at room temperature to measure the rheological properties. The viscometer was used to measure the PV, YP and GS of the drilling fluid at rotor speeds of 300 rpm and 600 rpm. The dial readings for both speeds are recorded as $\varnothing_{300}$ and $\varnothing_{600}$. For the GS, the drilling fluid was stirred at 600 rpm until it reached a steady dial reading value. The drilling fluid sample was then held for 10 s and the motor was stopped. The maximum reading was achieved and recorded as 10 s of GS; the same steps were repeated for 10 min of GS.

The filtration test was piloted by pouring the mud sample into the cell to within 1/2 inch of the top, and the filtrate was collected using a dry graduated cylinder placed under the drain tube. An OFITE filter press was used for this test. The system used N_2 to supply pressure and a standard filter paper. The pressure relief valve was opened and began to record the FL as a function of time. According to the API standard, the operating pressure was 100 psi and the temperature was atmospheric (77 °F). After 30 min, the FL was measured in cubic centimeters (to 0.1 ccs). The MCT was measured using a digital Vernier caliper, model Mitutoyo 500-197-20, to the nearest 1/32 inch [32,33].

3. Results and Discussion

3.1. Characterization of GO

The FTIR spectra of welding wastes and of GO derived from welding wastes are shown in Figure 2. The spectrum of welding wastes shows a weak absorption band at 3748 cm^{-1} corresponding to the O-H bond of Si-O-H. Bands appeared at 2986 cm^{-1}, showing the C-H bond, and at 1700, showing C=O; multiple bands appeared in the range of 1660 cm^{-1}–1541 cm^{-1}, corresponding to aromatic C=O configurations. Peaks at 1058 cm^{-1} and 955 cm^{-1} show an Si-O stretching vibration and Si-O-Al vibrations, whereas the bands appearing between 800 and 500 show metals-oxygen bonds [4,34]. These results show that the welding waste consists of silicates, aluminates and oxides of various metals, along with some proportion of graphite.

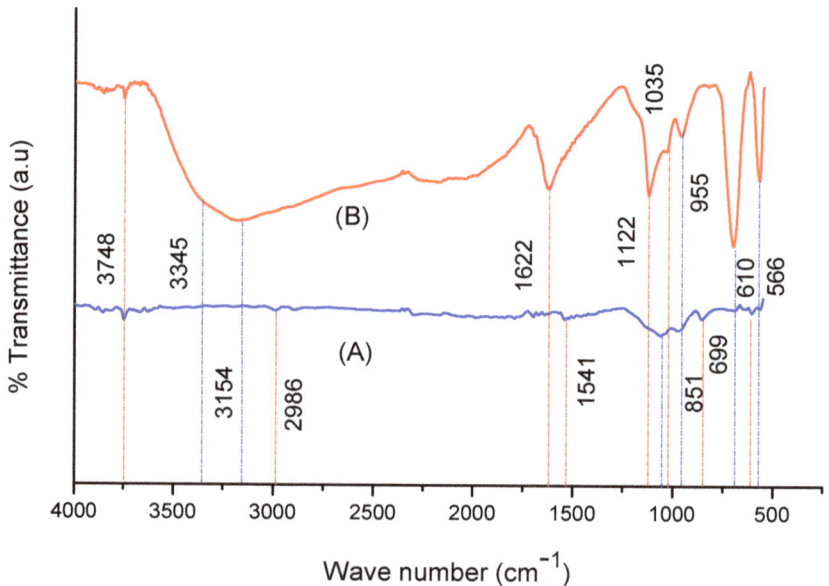

Figure 2. FTIR spectra of (**A**) welding wastes and (**B**) GO.

The FTIR spectrum of GO exhibits prominent bands positioned at 3748, 3345 and 3154 cm^{-1}, which correspond to the O-H stretching vibrations of Si-OH, O-H carboxylic acids and C-H aromatics [5,31]. The absorption bands positioned at 1622, 1122, 1035 and 955 cm^{-1} show C=C aromatics, carboxyl O=C-O, Si-O and Si-O-Al, respectively [5,35], whereas the FT-IR absorption bands appearing in the range of 610–500 cm^{-1}, correspond to various metal-oxygen linkages [6,7,36,37]. The results show that the GO sample also contains impurities such silica, alumina and several metal oxides.

The elemental composition of the welding-waste sample and GO prepared from welding waste was determined through EDX analysis. The EDX spectra of the samples and the % weight and atomic % values of various elements in the samples are displayed in Figure 3. These results reveal that the welding-waste sample contains various elements as their oxides, including Ti, Mn, Fe, Si, Ca, K, Na and Al, and of which Ti, Si, Fe and Mn are present as a high atomic % i.e., 13, 7, 4 and 3%, respectively. About 4 atomic % carbon in the sample is also present in the sample, the form of graphite. The welding electrode generally consists of a metal rod made of steel or wrought iron; the flux material around it contains cellulose, silica and oxides of various metals, such as Fe, Mn, Al, Ti, Ca and others [8,38]. During arc welding, the electrode transforms into residues of oxides. It was shown that the welding flux changes into granular powder during arc welding, and consists of alumina, silica, and oxides of Ca, Mn, Fe, Ti and other minerals [9,39]. The composition of welding wastes agrees with the composition of the components of the electrode. The elemental composition of the GO derived from welding wastes includes about 85% carbon and 7% oxygen, which confirms the synthesis of GO. It is clear from the results that Ti, Fe and Mn are not present in the sample where the other metals are present in very smaller quantities. It is inferred that during the synthesis of GO, the carbon content was enriched, and the other metals leached from the welding residue through acid treatment and the decantation processes. The presence of S in the GO sample is also an indication of contamination from the sulfuric treatment during the synthesis process.

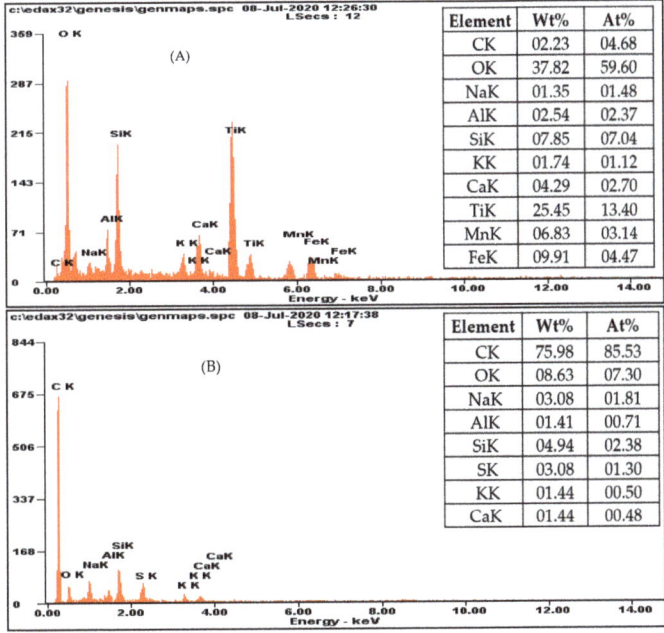

Figure 3. EDX profiles of (A) welding wastes and (B) GO.

The XRD pattern of the welding wastes and GO is given in Figure 4. The XRD pattern of welding wastes shows an amorphous hump and some sharp and intense peaks; the amorphous pattern appears at 2θ of 0 to 20° which indicates the presence of powder graphite. The high-intensity prominent peaks represent the presence of Fe_2O_3, MnO_2, TiO_2, silica and alumina as major components. The other less intense peaks indicate the presence of mullite and wollastonite in smaller proportions. The components of welding wastes exhibited by the XRD analysis agree with the literature reports [9,10,39,40].

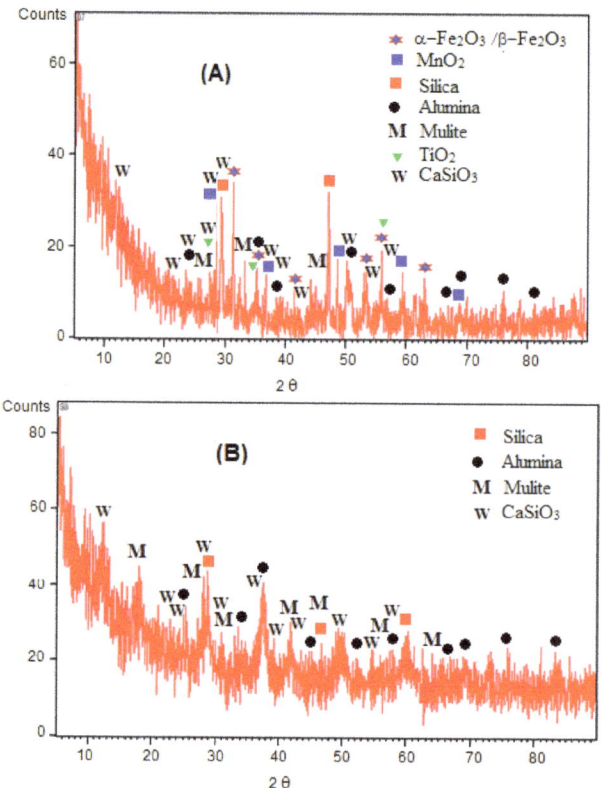

Figure 4. XRD patterns of (**A**) welding wastes and (**B**) GO samples.

In the case of GO, the XRD pattern shows a typical amorphous configuration of GO with a sharp peak at 10.5° 2θ [5,35]. Other configurations indicated by the XRD patterns of GO include silica, alumina, mullite and calcium silicate. As per EDX analysis, other minerals containing Na, S and K may also be present in the welding waste-derived GO. However, due to their small concentrations, the corresponding peaks do not appear in the XRD patterns.

The FESEM micrographs of welding wastes and GO derived from welding wastes are displayed in Figure 5 which illustrates that both samples exhibit an identical granular morphology. However, the granular size of the welding waste sample is non uniform, wherein the particles of different sizes ranging from nanosized grains to large lumps of two microns can be seen. Also, the grains seem agglomerated or glued together, which may be attributed to the presence of oxides of various metals. However, the GO sample represents regular-shaped, uniformed-sized particles with an average grain size of about half a micron.

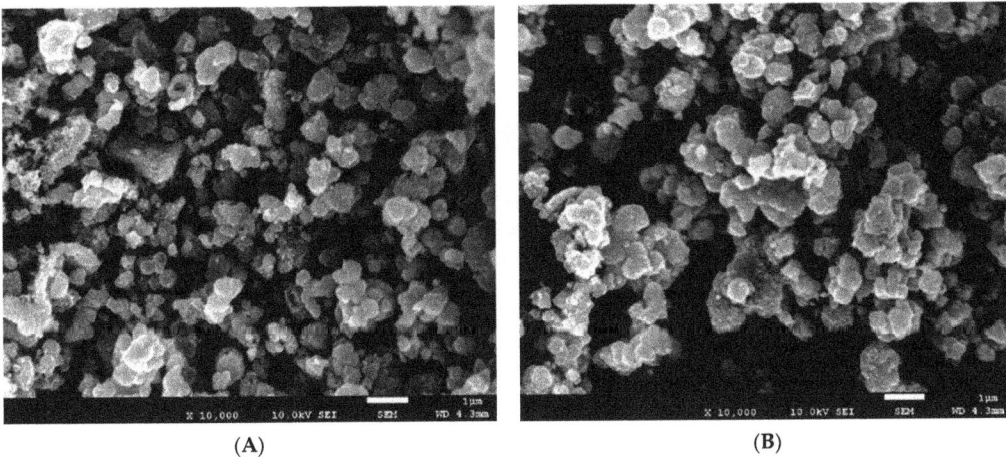

Figure 5. FESEM micrographs of (**A**) Welding waste sample and (**B**) GO.

3.2. Properties of Drilling Fluids

The rheological properties of drilling fluids such as PV, YP, 10 s and 10 min GS, API FL and MCT were determined as summarized in Tables 2–4 below.

Table 2. Rheological properties of commercial GO.

Concentration	Base WBDF	0.01 g	0.1 g	0.5 g
PV (cP)	14.0	16.0	16.5	17.0
AV (cP)	19.0	21.5	22.2	23.5
YP (lb/100 ft^2)	15.0	17.5	17.4	18.0
10sGS (lb/100 ft^2)	8.0	8.5	10.0	11.2
10mGS (lb/100 ft^2)	11.0	11.4	12.0	12.5
FL (mL)	9.0	7.4	7.0	6.8
MCT (mm)	2.82	1.52	1.44	1.13

Table 3. Rheological properties of raw welding waste.

Concentration	Base WBDF	0.01 g	0.1 g	0.5 g
PV (cP)	14.0	25.0	28.2	30.5
AV (cP)	19.0	28.0	30.4	31.4
YP (lb/100 ft^2)	15.0	15.0	16.2	17.2
10sGS (lb/100 ft^2)	8.0	7.0	8.0	8.5
10mGS (lb/100 ft^2)	11	9	10	10
FL (mL)	9.0	8.0	7.2	6.0
MCT (mm)	2.82	2.37	2.25	1.70

Table 4. Rheological properties of welding waste-derived GO.

Concentration	Base WBDF	0.01	0.1	0.5
PV (cP)	14.0	22.0	24.2	25.5
AV (cP)	19.0	24.0	26.4	27.0
YP (lb/100 ft^2)	15.0	17.0	18.2	19.0
10sGS (lb/100 ft^2)	8.0	6.0	6.5	7.5
10mGS (lb/100 ft^2)	11.0	8.0	9.5	9.7
FL (mL)	9.0	4.0	5.8	6.6
MCT (mm)	2.82	1.90	2.03	2.23

The rheological properties of WBDF with commercial GO as the additive are tabulated in Table 2.

The rheological properties of WBDF with raw welding waste as the additive are tabulated in Table 3.

The rheological properties of WBDF with welding waste-derived GO as the additive are tabulated in Table 4.

3.2.1. Plastic Viscosity

Figure 6 represents the comparison of plastic viscosity of raw welding waste, welding waste-derived GO and commercial GO-based drilling fluids with different concentrations. The results show that the PV of 0.5 ppb raw welding waste drilling fluid is 30.5 cP, which is the highest among them all. Drilling fluids with the presence of welding wastes increase dramatically compared with the base drilling mud and commercial GO drilling fluid. This might be due to the presence of the solid form welding waste, which increases the flow resistance of the drilling fluids [11,41]. PV increases when the mechanical friction between solids and liquids increases.

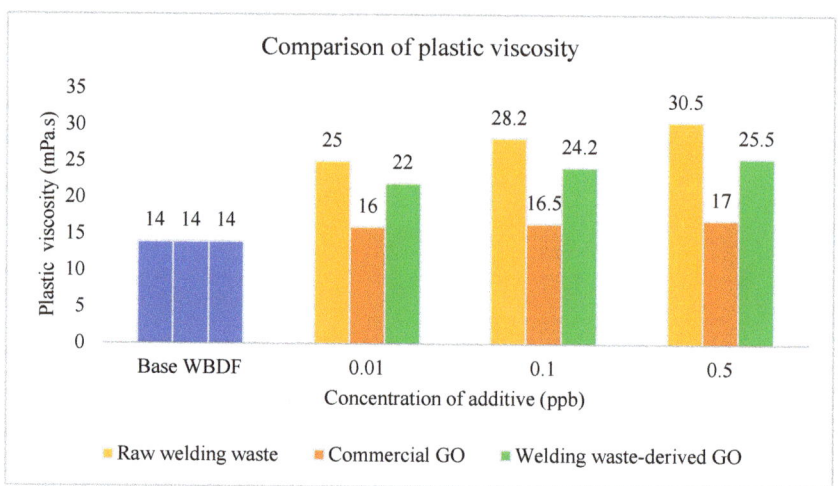

Figure 6. PV comparison of WBDF using different additives.

3.2.2. Yield Point

A comparison of YP for the different drilling fluid formulations at different concentrations is presented in Figure 7. The trend for all types of drilling fluids are that the YP increases with increasing concentration, except for 0.1 ppb of commercial GO. The lifting capacity of a high YP is better but would increase the cost of the power when the drilling fluid is pumped in the wellbore. Drilling fluids with high YP indicate that it has a higher suspension. The rock cuttings would not sink to the bottom when the drilling operation is stopped. Stuck pipes and lost circulation can be prevented if the YP of the drilling fluids are controlled. The YP of welding waste-derived GO drilling fluid increased from 15 lb/100 ft^2 to 19 lb/100 ft^2 when its concentration was 0.5 ppb. An increasing YP can improve the dynamic cutting suspension and efficiency of the hole cleaning of the drilling fluids [12,42].

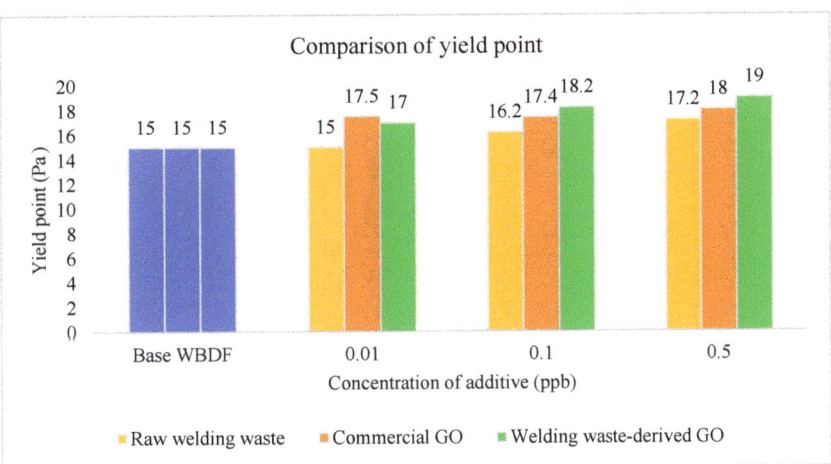

Figure 7. YP comparison of WBDF using different additives.

3.2.3. Gel Strength

GS is related to the viscosity of the drilling fluid and affects the ability of drilling fluids to lift rock cuttings [43]. Figures 8 and 9 show that GS increased with the increasing concentration of additives when different types of additives were added into WBDF. The 10 s GS and 10 min GS of commercial GO based drilling fluid increased to 11.2 Ib/100 ft^2 and 12.5 Ib/100 ft^2, respectively, when the concentration of additives was 0.5 ppb. Drilling fluids have higher suspension power with a higher GS when the drilling operation is stopped. For the 10 s GS in Figure 8, the GS deceased slightly when raw welding waste and welding waste-derived GO were added into WBDF. From Figure 9, it can be seen that the 10 min GS increased gradually as the concentration of additives increased.

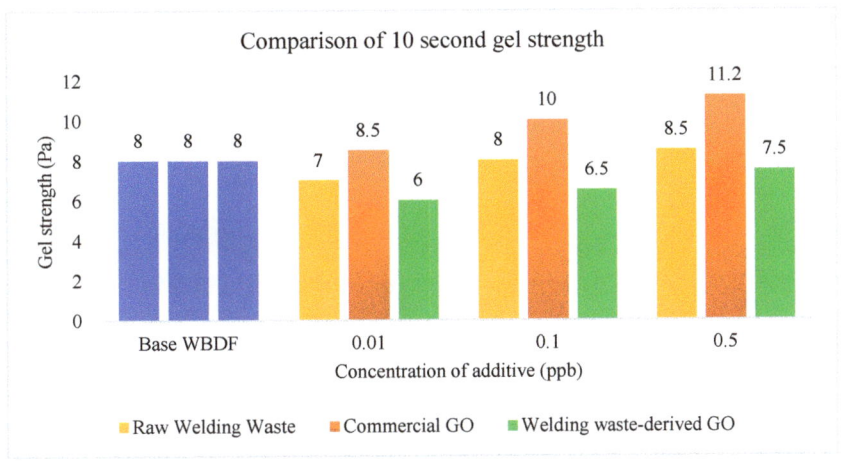

Figure 8. 10 s GS comparison of WBDF using different additives.

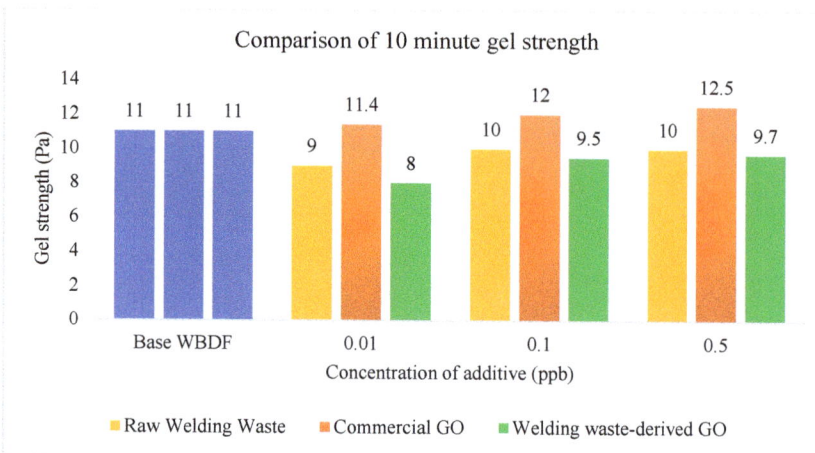

Figure 9. 10 min GS comparison of WBDF using different additives.

3.2.4. Filtrate Loss

The FL of the three types of drilling fluid at the different concentrations of 0.01 ppb, 0.1 ppb and 0.5 ppb is shown in Figure 10. The base fluid displayed a filtrate loss volume of 9 mL. The fluid loss was significantly reduced to 6.6, 5.8 and 4 mL after the addition of 0.5, 0.1 and 0.01 wt% of GO nanoparticles to the WBDF, respectively, as shown in Figure 9. Raw welding waste-derived GO drilling fluid with 0.01 ppb shows the best result as the FL decreased from 9 mL to 4 mL—the highest reduction. The GO nanoparticles remarkably minimize the permeability and porosity of the mud cake [44]. By consolidating the attractive force between the particles, the volume of FL is reduced. On the other hand, when the amount of welding waste-derived GO was increased, the FL also increased.

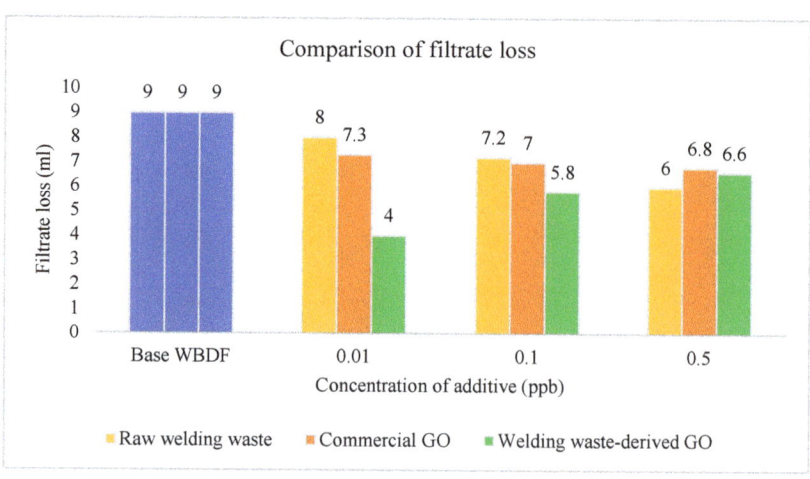

Figure 10. FL comparison of WBDF using different additives.

The usage of GO as a drilling fluid additive is likely because it has a special surface area of 2391 m^2/g [13]. This acts as a solid penetrating layer that can make the formation of mud cake in drilling mud stronger, thereby preventing the mud from flooding into the formation. Furthermore, GO possesses another interesting characteristic, namely separation of the solution and the formation of a strong paper-shaped material capable of efficiently preventing drilling mud leaks into the well wall. For instance, dispersed GO flakes can be filtered out of the solution and pressed to make a strong paper-like material, which results from the robust tile-like interlocking of the flakes. This could be beneficial for making a thin impermeable film to prevent FL in the wellbore [23,45].

It is believed that FL is affected by the concentration of nanomaterials added after a specific point as 0.01 ppb is the optimum concentration of waste-derived GO drilling fluid with the lowest FL. Excessive nanomaterials added to drilling fluid may reduce the effectiveness, make it costlier and increase the chance of formation damage [46,47].

3.2.5. Mud Cake Thickness

The optimum thickness of mud cake helps to increase the stability of the wellbore and decrease mud invasion; hence, thin mud cake is recommended [48]. This is due to thick filter cake increasing the chance of the drill pipe becoming stuck when in contact with the mud cake under pressure. Figure 11 shows the thickness of mud cake with the addition of different amounts of raw welding waste, commercial GO and welding waste-derived GO. The effectiveness of the drilling fluid is closely related to the permeability and porosity of the mud cake. As shown in Figure 11, a decrease in fluid loss volume results in a decrease in MCT. It is clear that by adding the lowest GO concentration (0.01 wt%) to WBDF, the lowest FL volume was established, and the thinnest mud cake was obtained, demonstrating significant enhancement by lowering the FL by 55.6% when compared with basic WBDF. Similarly, MCT was reduced to 32.6% using welding waste-derived GO. As a result, GO nanoparticles were critical in blocking the nanopores in the filter cake made of bentonite particles. For commercial GO and raw welding waste-derived GO, significant results are obtained, as MCT is reduced with a minimum amount of additive. However, large graphene-cut stacks in the water medium experience issues with WBDF. Therefore, GO, which is more water resistant and possesses the same layered morphology, is capable of forming the desired mud cake. Since GO sheets are well exfoliated, they could be added at substantially lower concentrations than clay-based additives to obtain the desired performance. More prominently, the nanometer thickness of the GO flakes could also result in much thinner filter cakes than those obtained using clay-based materials. The thickness of a filter cake is directly correlated with the differential torque needed to rotate the pipe during drilling operations, to the drilling time and to drilling costs. In this work, GO is further appealing as it offers the prospect of a waste-derived mechanism and inexpensive technology [20,22,45].

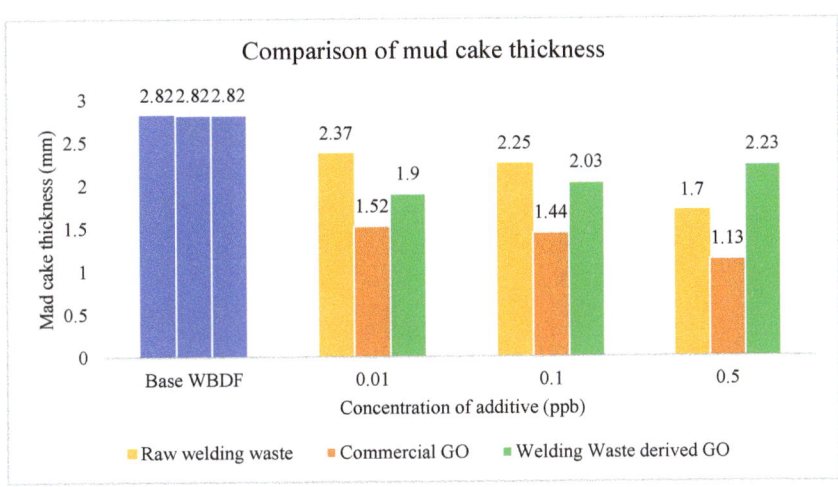

Figure 11. MCT comparison of WBDF using different additives.

Both additives were able to form the thinner mud cake compared with the conventional WBDF. From this study, it can be concluded that waste-derived GO can prominently improve the filtration properties of WBDF [49,50].

4. Conclusions

In this work, unwanted welding waste-derived GO was prepared in a novel way using a modified Hummers' method to improve the filtration properties of WBDF. A comparison of three additives including raw welding waste, welding waste-derived GO and commercial GO was conducted, using three different concentrations i.e., low (0.01 wt%), medium (0.1 wt%) and high (0.5 wt%) under API standard conditions. The following conclusions can be drawn based on this research;

- The effect of the lowest concentration of welding waste-derived GO showed the lowest FL up to 55.6%.
- The remarkable reduction of MCT was obtained by adding highest concentration of raw welding waste (39.7%). However, the highest concentration of welding waste-derived GO and commercial GO showed reductions of 20.9% and 59.9%, respectively.
- The lowest concentration of commercial GO showed an MCT reduction of 46.1% compared with welding waste-derived GO, which showed a reduction of up to 32.6%. The results verify that the use of novel nanocomposites in drilling fluids is capable of increasing efficiency, while reducing operating cost. These waste-derived economical additives can be excellent alternatives to commercial GO.
- The linked structure of GO allows more water to be trapped between layers. This causes an increase in viscosity and yield stress, while reducing fluid filtration.
- The addition of waste-derived GO could be a useful substitute of commercial GO.
- The major cost to produce GO is the cost of chemicals, equipment and labor. The chemicals required include H_2SO_4, $KMnO_4$, $NaNO_3$ and H_2O_2, which are all abundantly available in the open market at very low prices. Moreover, these are commonly used chemicals and are available in almost every lab. The cost of these consumables is same as for commercial GO and GO derived from welding wastes. These only differ in the supply of raw materials. For commercial GO, the raw material is pure graphite, which must be imported at high cost, whereas for synthesis of GO from welding waste, the raw material is a waste product and available free of cost. Thus, the total production cost of welding waste-derived GO is much lower than that of t commercial GO.
- This research focused on the conversion of hazardous waste into GO which validates the conversion of many other waste materials into graphene derivatives and potential filtrate loss agents. Thus, this study paves the way for utilizing a variety of waste sources in sustainable and cost-effective drilling operations. The availability of carbonaceous industrial waste, therefore, can stimulate initiatives to convert such industrial waste into value-added products.

5. Future Perspectives

This work supports the view that the effectiveness of drilling fluids containing different type of additives is significantly increased. When the conventional formulations for welding waste-derived additive-based formulations were substituted and the optimum concentration of additives was added, FL and MCT were reduced. The results prove that the novel welding waste-derived GO-based drilling fluids are capable of increasing efficiency, while reducing operating costs. This presents a golden opportunity to extract graphene from other industrial wastes and utilize it to further enhance drilling fluid properties. Welding wastes consist of toxic substances such as aluminum, titanium oxide and others. Metal oxides are hazardous to the environment and marine life. Hence, reuse and extraction into valuable products from raw welding waste is very important. In conclusion, lower amounts of toxins will be released into the environment, leading to a lower impact on the atmosphere.

Nanotechnology recently introduced new formulations of drilling fluids to the oil and gas industry. This technology can rearrange the properties of nanomaterials to produce more attractive properties which are essential in drilling fluids. By-products from the construction industry such as welding wastes can be reused in WBDFs and can serve as low-cost additives compared with commercial products in terms of optimizing the performance of conventional drilling fluids. The development of advanced methods is, therefore, required for the production of nanomaterials on a large scale and with cost-effective strategies for commercialization. In addition, proper guidelines for the handling and storage of nanomaterials must be implemented. This is to prevent contamination of the materials which leads to high surface activity and degrades their effectiveness.

Finally, novel routes are required to be discovered for the extraction of GO from waste materials. The use of welding waste-derived GO improves the rheological properties of drilling fluids. Researchers from different fields of expertise should use their skills and experience to improve technology and create new remarkable materials such as green nanocomposites. Other parameters such as high temperature and high pressure can also be used in future studies to investigate the effect of aging on the drilling fluids after the addition of novel additives.

Author Contributions: R.I. and W.A., original concept and initial draft of the paper; R.I., W.A., A.S., G.K. and M.K., processing, data analysis and validation; B.M.J. supervised and coordinated the work; G.K. and R.I., funding acquisition. All authors have read and agreed to the published version of the manuscript.

Funding: The authors would like to thank the Malaysia-Thailand Joint Authority (MTJA) under grant number IF062-2019, the Fundamental Research Grant Scheme FP050-2019A and the RU Geran-Fakulti Program GPF054B-2020 at the University of Malaya for providing funds during the course of this study. This work was also supported by the National Priorities Research Program Grant No. NPRP11S-1128-170042 from the Qatar National Research Fund (member of The Qatar Foundation).

Institutional Review Board Statement: Not Applicable.

Informed Consent Statement: Not Applicable.

Data Availability Statement: Not Applicable.

Conflicts of Interest: The authors declare that there are no conflict of interest regarding the publication of this manuscript.

Abbreviations

API	American Petroleum Institute
AV	Apparent viscosity
EDX	Energy Dispersive X-ray analysis
PV	Plastic viscosity
YP	Yield point
GS	Gel strength
FL	Filtrate loss
FTIR	Fourier Transform Infrared Spectroscopy
FESEM	Field Emission Scanning Electron Microscopy
MCT	Mud cake thickness
GO	Graphene oxide
WBDF	Water-based drilling fluids
XRD	X-ray Diffraction Crystallography

References

1. Saini, S.; Singh, K. Utilisation of steel slag as welding flux and its chemical and thermal characterisation. *Mater. Sci. Technol.* **2022**, 1–12. [CrossRef]
2. Garg, J.; Singh, K. Slag recycling in submerged arc welding and its effects on the quality of stainless steel claddings. *Mater. Des.* **2016**, *108*, 689–698. [CrossRef]

3. Kaptanoglu, M.; Eroglu, M. Reusability of metallurgical slag waste in submerged arc welding powder for hardfacing. *Mater. Test.* **2020**, *62*, 177–183. [CrossRef]
4. Saleem, H.; Zaidi, S.J.; Alnuaimi, N.A. Recent Advancements in the Nanomaterial Application in Concrete and Its Ecological Impact. *Materials* **2021**, *14*, 6387. [CrossRef]
5. Kaya, M. Recovery of metals and nonmetals from electronic waste by physical and chemical recycling processes. *Waste Manag.* **2016**, *57*, 64–90. [CrossRef]
6. Ramesh, S.T.; Gandhimathi, R.; Nidheesh, P.V.; Rajakumar, S.; Prateepkumar, S. Use of furnace slag and welding slag as replacement for sand in concrete. *Int. J. Energy Environ. Eng.* **2013**, *4*, 1–6. [CrossRef]
7. Loaiza, A.; Colorado, H.A. Marshall stability and flow tests for asphalt concrete containing electric arc furnace dust waste with high ZnO contents from the steel making process. *Constr. Build. Mater.* **2018**, *166*, 769–778. [CrossRef]
8. Laska, A.; Szkodo, M. Manufacturing parameters, materials, and welds properties of butt friction stir welded joints–overview. *Materials* **2020**, *13*, 4940. [CrossRef]
9. Magzoub, M.I.; Ibrahim, M.H.; Nasser, M.S.; El-Naas, M.H.; Amani, M. Utilization of Steel-Making Dust in Drilling Fluids Formulations. *Processes* **2020**, *8*, 538. [CrossRef]
10. Bayat, A.E.; Moghanloo, P.J.; Piroozian, A.; Rafati, R. Experimental investigation of rheological and filtration properties of water-based drilling fluids in presence of various nanoparticles. *Colloids Surf. A Physicochem. Eng. Asp.* **2018**, *555*, 256–263. [CrossRef]
11. Misbah, B.; Sedaghat, A.; Rashidi, M.; Sabati, M.; Vaidyan, K.; Ali, N.; Omar, M.A.A.; Dehshiri, S.S.H. Friction reduction of Al2O3, SiO2, and TiO2 nanoparticles added to non-Newtonian water based mud in a rotating medium. *J. Pet. Sci. Eng.* **2022**, *217*, 110927. [CrossRef]
12. Lin, L.; Feng, K.; Wang, P.; Wan, Z.; Kong, X.; Li, J. Hazardous waste from the global shipbreaking industry: Historical inventory and future pathways. *Glob. Environ. Change* **2022**, *76*, 102581. [CrossRef]
13. Binnemans, K.; Jones, P.T.; Blanpain, B.; Van Gerven, T.; Pontikes, Y. Towards zero-waste valorisation of rare-earth-containing industrial process residues: A critical review. *J. Clean. Prod.* **2015**, *99*, 17–38. [CrossRef]
14. Mokshein, S.E. Welding Waste In Vocational College: Should It Be Properly Managed? *J. Teknol.* **2015**, *72*. [CrossRef]
15. Rizkya, I.; Syahputri, K.; Sari, R.M.; Siregar, I. 5S implementation in welding workshop—A lean tool in waste minimization. In *IOP Conference Series: Materials Science and Engineering*; IOP Publishing: Bristol, UK, 2019; Volume 505, p. 012018.
16. Ikram, R.; Jan, B.M.; Ahmad, W. An overview of industrial scalable production of graphene oxide and analytical approaches for synthesis and characterization. *J. Mater. Res. Technol.* **2020**, *9*, 11587–11610. [CrossRef]
17. Ikram, R.; Mohamed Jan, B.; Vejpravova, J.; Choudhary, M.I.; Zaman Chowdhury, Z. Recent advances of graphene-derived nanocomposites in water-based drilling fluids. *Nanomaterials* **2020**, *10*, 2004. [CrossRef]
18. Villada, Y.; Busatto, C.; Casis, N.; Estenoz, D. Use of synthetic calcium carbonate particles as an additive in water-based drilling fluids. *Colloids Surf. A Physicochem. Eng. Asp.* **2022**, *652*, 129801. [CrossRef]
19. Kusrini, E.; Oktavianto, F.; Usman, A.; Mawarni, D.P.; Alhamid, M.I. Synthesis, characterization, and performance of graphene oxide and phosphorylated graphene oxide as additive in water-based drilling fluids. *Appl. Surf. Sci.* **2020**, *506*, 145005. [CrossRef]
20. Ikram, R.; Mohamed Jan, B.; Sidek, A.; Kenanakis, G. Utilization of eco-friendly waste generated nanomaterials in water-based drilling fluids; state of the art review. *Materials* **2021**, *14*, 4171. [CrossRef]
21. Ma, J.; Pang, S.; Zhang, Z.; Xia, B.; An, Y. Experimental study on the polymer/graphene oxide composite as a fluid loss agent for water-based drilling fluids. *ACS Omega* **2021**, *6*, 9750–9763. [CrossRef]
22. Ikram, R.; Jan, B.M.; Vejpravova, J. Towards recent tendencies in drilling fluids: Application of carbon-based nanomaterials. *J. Mater. Res. Technol.* **2021**, *15*, 3733–3758. [CrossRef]
23. Rafati, R.; Smith, S.R.; Haddad, A.S.; Novara, R.; Hamidi, H. Effect of nanoparticles on the modifications of drilling fluids properties: A review of recent advances. *J. Pet. Sci. Eng.* **2018**, *161*, 61–76. [CrossRef]
24. Zhao, X.; Qiu, Z.; Sun, B.; Liu, S.; Xing, X.; Wang, M. Formation damage mechanisms associated with drilling and completion fluids for deepwater reservoirs. *J. Pet. Sci. Eng.* **2019**, *173*, 112–121. [CrossRef]
25. Vryzas, Z.; Kelessidis, V.C. Nano-based drilling fluids: A review. *Energies* **2017**, *10*, 540. [CrossRef]
26. Martín-Alfonso, M.J.; Pozo, J.; Delgado-Sánchez, C.; Martínez-Boza, F.J. Thermal and Rheological Properties of Hydrophobic Nanosilica in Sunflower Oil Suspensions at High Pressures. *Nanomaterials* **2021**, *11*, 3037. [CrossRef]
27. Wang, Q.; Slaný, M.; Gu, X.; Miao, Z.; Du, W.; Zhang, J.; Gang, C. Lubricity and rheological properties of highly dispersed graphite in clay-water-based drilling fluids. *Materials* **2022**, *15*, 1083. [CrossRef]
28. Zinchik, S.; Klinger, J.L.; Westover, T.L.; Donepudi, Y.; Hernandez, S.; Naber, J.D.; Bar-Ziv, E. Evaluation of fast pyrolysis feedstock conversion with a mixing paddle reactor. *Fuel Process. Technol.* **2018**, *171*, 124–132. [CrossRef]
29. Ikram, R.; Jan, B.M.; Ahmad, W. Advances in synthesis of graphene derivatives using industrial wastes precursors; prospects and challenges. *J. Mater. Res. Technol.* **2020**, *9*, 15924–15951. [CrossRef]
30. Chen, J.; Yao, B.; Li, C.; Shi, G. An improved Hummers method for eco-friendly synthesis of graphene oxide. *Carbon* **2013**, *64*, 225–229. [CrossRef]
31. Yu, H.; Zhang, B.; Bulin, C.; Li, R.; Xing, R. High-efficient synthesis of graphene oxide based on improved hummers method. *Sci. Rep.* **2016**, *6*, 1–7. [CrossRef]

32. Boudouh, D.; Ikram, R.; Mohamed Jan, B.; Simon Cornelis Metselaar, H.; Hamana, D.; Kenanakis, G. Synthesis, Characterization and Filtration Properties of Ecofriendly Fe_3O_4 Nanoparticles Derived from Olive Leaves Extract. *Materials* **2021**, *14*, 4306. [CrossRef] [PubMed]
33. Yang, J.; Sun, J.; Wang, R.; Liu, F.; Wang, J.; Qu, Y.; Wang, P.; Huang, H.; Liu, L.; Zhao, Z. Laponite-polymer composite as a rheology modifier and filtration loss reducer for water-based drilling fluids at high temperature. *Colloids Surf. A Physicochem. Eng. Asp.* **2022**, *655*, 130261. [CrossRef]
34. Cortea, I.M.; Ghervase, L.; Rădvan, R.; Serițan, G. Assessment of Easily Accessible Spectroscopic Techniques Coupled with Multivariate Analysis for the Qualitative Characterization and Differentiation of Earth Pigments of Various Provenance. *Minerals* **2022**, *12*, 755. [CrossRef]
35. Luo, X.; Wang, C.; Luo, S.; Dong, R.; Tu, X.; Zeng, G. Adsorption of As (III) and As (V) from water using magnetite Fe3O4-reduced graphite oxide–MnO_2 nanocomposites. *Chem. Eng. J.* **2012**, *187*, 45–52. [CrossRef]
36. Chandra, V.; Park, J.; Chun, Y.; Lee, J.W.; Hwang, I.C.; Kim, K.S. Water-dispersible magnetite-reduced graphene oxide composites for arsenic removal. *ACS Nano* **2010**, *4*, 3979–3986. [CrossRef]
37. Chumming, J.; Xiangqin, L. Electrochemical synthesis of Fe3O4-PB nanoparticles with core-shell structure and its electrocatalytic reduction toward H2O2. *J. Solid State Electrochem.* **2009**, *13*, 1273–1278. [CrossRef]
38. Mbah, C.N.; Ugwuanyi, B.C.; Nnakwo, K.C. Effect of welding parameters on the mechanical properties of arc–welded C1035 medium carbon steel. *Int. J. Res. Adv. Eng. Technol.* **2019**, *5*, 30–33.
39. Francis, A.A.; Abdel Rahman, M.K. Transforming submerged-arc welding slags into magnetic glass-ceramics. *Int. J. Sustain. Eng.* **2016**, *9*, 411–418. [CrossRef]
40. Ananthi, A.; Karthikeyan, J. Properties of industrial slag as fine aggregate in concrete. *Int. J. Eng. Technol. Innov.* **2015**, *5*, 132.
41. Chu, Q.; Lin, L. Synthesis and properties of an improved agent with restricted viscosity and shearing strength in water-based drilling fluid. *J. Pet. Sci. Eng.* **2019**, *173*, 1254–1263. [CrossRef]
42. Moraveji, M.K.; Ghaffarkhah, A.; Agin, F.; Talebkeikhah, M.; Jahanshahi, A.; Kalantar, A.; Amirhosseini, S.F.; Karimifard, M.; Mortazavipour, S.I.; Sehat, A.A.; et al. Application of amorphous silica nanoparticles in improving the rheological properties, filtration and shale stability of glycol-based drilling fluids. *Int. Commun. Heat Mass Transf.* **2020**, *115*, 104625. [CrossRef]
43. Saboori, R.; Sabbaghi, S.; Kalantariasl, A. Improvement of rheological, filtration and thermal conductivity of bentonite drilling fluid using copper oxide/polyacrylamide nanocomposite. *Powder Technol.* **2019**, *353*, 257–266. [CrossRef]
44. Huang, W.A.; Wang, J.W.; Lei, M.; Li, G.R.; Duan, Z.F.; Li, Z.J.; Yu, S.F. Investigation of regulating rheological properties of water-based drilling fluids by ultrasound. *Pet. Sci.* **2021**, *18*, 1698–1708. [CrossRef]
45. Kosynkin, D.V.; Ceriotti, G.; Wilson, K.C.; Lomeda, J.R.; Scorsone, J.T.; Patel, A.D.; Friedheim, J.E.; Tour, J.M. Graphene oxide as a high-performance fluid-loss-control additive in water-based drilling fluids. *ACS Appl. Mater. Interfaces* **2012**, *4*, 222–227. [CrossRef] [PubMed]
46. Kumar, B.D. and Kulkarni, S.D. A review of synthetic polymers as filtration control additives for water-based drilling fluids for high-temperature applications. *J. Pet. Sci. Eng.* **2022**, *215*, 110712.
47. Zhong, H.; Guan, Y.; Su, J.; Zhang, X.; Lu, M.; Qiu, Z.; Huang, W. Hydrothermal synthesis of bentonite carbon composites for ultra-high temperature filtration control in water-based drilling fluid. *Appl. Clay Sci.* **2022**, *230*, 106699. [CrossRef]
48. Li, J.; Sun, J.; Lv, K.; Ji, Y.; Liu, J.; Huang, X.; Bai, Y.; Wang, J.; Jin, J.; Shi, S. Temperature-and Salt-Resistant Micro-Crosslinked Polyampholyte Gel as Fluid-Loss Additive for Water-Based Drilling Fluids. *Gels* **2022**, *8*, 289. [CrossRef] [PubMed]
49. Gamal, H.; Elkatatny, S.; Basfar, S.; Al-Majed, A. Effect of pH on rheological and filtration properties of water-based drilling fluid based on bentonite. *Sustainability* **2019**, *11*, 6714. [CrossRef]
50. Medved, I.; Gaurina-Međimurec, N.; Novak Mavar, K.; Mijić, P. Waste mandarin peel as an eco-friendly water-based drilling fluid additive. *Energies* **2022**, *15*, 2591. [CrossRef]

Article

Plasmon Damping Rates in Coulomb-Coupled 2D Layers in a Heterostructure

Dipendra Dahal [1,*], Godfrey Gumbs [2], Andrii Iurov [3] and Chin-Sen Ting [1]

[1] Texas Center for Superconductivity and Department of Physics, University of Houston, Houston, TX 77204, USA
[2] Department of Physics and Astronomy, Hunter College, City University of New York, 695 Park Avenue, New York, NY 10065, USA
[3] Department of Physics and Computer Science, Medgar Evers College, City University of New York, Brooklyn, NY 11225, USA
* Correspondence: hn6565@wayne.edu

Abstract: The Coulomb excitations of charge density oscillation are calculated for a double-layer heterostructure. Specifically, we consider two-dimensional (2D) layers of silicene and graphene on a substrate. From the obtained surface response function, we calculated the plasmon dispersion relations, which demonstrate how the Coulomb interaction renormalizes the plasmon frequencies. Most importantly, we have conducted a thorough investigation of how the decay rates of the plasmons in these heterostructures are affected by the Coulomb coupling between different types of two-dimensional materials whose separations could be varied. A novel effect of nullification of the silicene band gap is noticed when graphene is introduced into the system. To utilize these effects for experimental and industrial purposes, graphical results for the different parameters are presented.

Keywords: plasmon; graphene; silicene; heterostructure

Citation: Dahal, D.; Gumbs, G.; Iurov, A.; Ting, C.-S. Plasmon Damping Rates in Coulomb-Coupled 2D Layers in a Heterostructure. *Materials* 2022, 15, 7964. https://doi.org/10.3390/ma15227964

Academic Editors: Victoria Samanidou and Eleni Deliyanni

Received: 9 October 2022
Accepted: 2 November 2022
Published: 11 November 2022

Copyright: © 2022 by the authors. Licensee MDPI, Basel, Switzerland. This article is an open access article distributed under the terms and conditions of the Creative Commons Attribution (CC BY) license (https://creativecommons.org/licenses/by/4.0/).

1. Introduction

A huge number of researchers from various disciplines have been showing their interest in new materials, silicene especially, after the development of its fabrication process in 2012 [1]. Because of its exceptional potential applications in electronic and optoelectronic devices, many industries are making substantial investments to harness its properties. Additionally, before making investments for commercial gain, both theoreticians and experimentalists have been exploring this material for many years. A credit of foremost importance goes to Takeda and Shiraishi [2], who, in 1994, dealt with the atomic and electronic structure of the material for the first time. These authors calculated the band structure of silicon in the corrugated stage having optimized atomic geometry. This work, though very novel, did not receive the attention it deserves until 2004, when single-layer carbon atoms named graphene were fabricated in the laboratory from graphite by Novoselov et al. [3]. Their research not only validated the stability of two-dimensional (2D) material but also opened the door for new research on thin film materials, silicene being one of them.

Both silicene and graphene were studied in parallel. The former has a buckled crystal geometry, whereas the latter has a honeycomb planar geometry. Due to this, differences arise between them. Ab initio calculations showed that the bandgap of silicene is electrically tunable [4–6], which is an advantageous property for designing a field effect transistor that works at room temperature. Another distinct difference between these two materials is the strength of the spin-orbital coupling (SOC), which is very weak in graphene. Consequently, the quantum spin Hall effect occurs at extremely low temperatures [7,8]. In contrast to this, silicene displays quantum spin Hall effect at temperature 18 K, far higher than that for graphene.

Several investigations have been carried out on both graphene and silicene with respect to transport phenomena [9–16], as well as their magnetic and electric field effects [3,17–23], the fabrication process [24–27], plasmonic behavior [28–38] and the doping effect on layered graphene and graphene-like heterostructure systems [39,40].

Recently, Dong et al. [41] studied the plasmonic behavior and its decay mode in multilayer graphene structure, however, an extensive literature search on the plasmon-related studies indicates that there has been no investigation regarding plasmon excitations and their decay rates due to Landau damping for composite silicene and graphene materials. This hybrid material could have significant benefits for use in the advancement of quantum information technology [42–44], sensing devices [45–47] and protein analytic clinical devices, [47–50]. Based on these impressive potential applications, we are motivated, in this work, to choose a system composed of silicene and graphene accompanied by a conducting substrate. A detailed review of the plasmon properties in graphene and various graphene-based structures has been presented in Ref. [51].

The plasmon mode is tunable by the thickness of the substrate and the variation of material behavior. We first determine the surface response function of the structure, the same technique used recently by Gumbs et al. [52,53], which gives us the condition for the existence of the plasmon dispersion. The analytical result for the surface response function is further used for different limiting cases and a comprehensive comparison is made with a variety of structures composed of different graphene–silicene compositions. Furthermore, the same function is used to obtain the Landau damping rate of the plasmon modes whose numerical calculation demonstrates that its variation depends on the layer separation, types of dielectric used and the type of 2D layer employed.

We have organized the rest of our paper as follows. In Section 2, we present the core idea of our work where we show the analytical result for the surface response function for the chosen structure. Under limiting conditions, the result is used to derive the results for a variety of conditions. The graphical results and their interpretation are presented in Section 3. We conclude our paper with a summary of our main results and conclusions in Section 4.

2. Theory

In this section, we analyze the plasmonic behavior of the heterostructure consisting of graphene and silicene together for which we employ the low-energy form of the Hamiltonian near the K point in the Brillouin zone. One significant difference between the Hamiltonian of graphene and silicene is that a small band gap, Δ is present in the silicene energy band structure, which is due to spin-orbit coupling and applied external electric field. This band gap is not seen in intrinsic graphene.

2.1. Silicene

We now briefly describe the case pertaining to silicene whose Hamiltonian in the continuum limit is given by:

$$H_\xi = \hbar v_F(\xi k_x \hat{\tau}_x + k_y \hat{\tau}_y) - \xi \Delta_{so} \sigma_z \tau_z + \Delta_z \hat{\tau}_z , \quad (1)$$

where $\hat{\tau}_{x,y,z}$ and $\sigma_{x,y,z}$ are Pauli matrices corresponding to two spin and coordinate subspaces, $\xi = \pm 1$ for the K and K' valleys, $v_F (\approx 5 \times 10^5)$ m/s is the Fermi velocity for silicene [5,54], k_x and k_y are the wave vector components measured relative to the K points. The first term represents the low-energy Hamiltonian, whereas the second term denotes the Kane–Mele system [7] for intrinsic spin-orbit coupling with an associated spin-orbit band gap of Δ_{so} (of order 5 to 30 meV can could reach up to 100 meV) [55]. The last term in the expression describes the sublattice potential difference that arises from the application of a perpendicular electric field. Equation (1) for the Hamiltonian is a block diagonal in 2×2 matrices labeled by valley (ξ) and spin $\sigma = \pm 1$ for up and down spin, respectively. These matrices are given by [55]:

$$\hat{H}_{\sigma\xi} = \begin{pmatrix} -\sigma\xi\Delta_{so} + \Delta_z & \hbar v_F(\xi k_x - ik_y) \\ \hbar v_F(\xi k_x + ik_y) & \sigma\xi\Delta_{so} - \Delta_z \end{pmatrix}. \quad (2)$$

This gives the low-energy eigenvalues as:

$$E_k = \pm\sqrt{\hbar^2 v_f^2 |k|^2 + \Delta_{\xi\sigma}^2} \qquad (3)$$

where $\Delta_{\sigma\xi} = |\sigma\xi\Delta_{so} - \Delta_z|$.

2.2. Graphene

The low-energy model Hamiltonian for monolayer graphene is similar to that in Equation (2) with the diagonal terms replaced by zero and ξ labeling the valley. In this regime, the Hamiltonian for intrinsic graphene is given by [56]:

$$\hat{H} = \hbar v_F \begin{pmatrix} 0 & (\xi k_x - ik_y) \\ (\xi k_x + ik_y) & 0 \end{pmatrix} \qquad (4)$$

with the linear energy dispersion, $E_k = \pm\hbar v_f |k|$ in either valley.

2.3. Polarization Function: $\Pi(q,\omega)$

Considerable work has been completed on the dynamical properties involving the use of the dielectric function $\epsilon(q,\omega)$ of various types of free-standing 2D systems [57–59] under different conditions. These include temperature effects [60,61], the role of an ambient magnetic field for the 2D electron gas (2DEG), graphene, silicene [62] and the dice lattice [58]. For a single 2D layer, one can extract the plasmon dispersion relation and damping rate by employing the dielectric function. However, the situation is more complicated for a multi-layer heterostructure that relies on knowledge of the surface response function, which we have presented in detail below. However, in either case, we need to calculate the polarization function obtained in the random phase approximation (RPA). For a 2D layer surrounded by a medium with dielectric constant ϵ_b, the dynamic dielectric function is given by:

$$\epsilon(q,\omega) = 1 - V(q)\Pi(q,\omega), \qquad (5)$$

where $V(q) = \frac{2\pi e^2}{4\pi\epsilon_0\epsilon_b q}$ is the Coulomb interaction potential and ϵ_0 is the permittivity of free space, q is the wave vector and e is the electron charge. The polarization function is an important quantity in calculations of the transport, collective charge motion and charge screening properties of the material. In the one-loop approximation, the polarization function for gapped graphene is given by [63]:

$$\Pi^0(q,\omega) = \int \frac{d^2 k}{2\pi^2} \sum_{s,s'=\pm 1} \left\{ \frac{\hbar^2 v_f^2 (\mathbf{k+q})\cdot \mathbf{k} + \Delta_{\sigma,\xi}^2}{E_k \cdot E_{|k+q|}} \right\} \frac{f_0(sE_k - E_F, T) - f_0(s'E_{|k+q|} - E_F, T)}{sE_k - s'E_{|k+q|} - \hbar(\omega + i\delta)} \qquad (6)$$

where the angle between $\mathbf{k}$ and $\mathbf{k+q}$ is $\theta_{k,k+q}$. At zero temperature, the Fermi function $f_0(z)$ is just a step function. The analytical expression for the polarization function for the silicene and graphene monolayer is given by Tabert et al. [62] and Wunsch et al. [57], respectively.

We now turn our attention to a crucial consideration in this paper regarding the structure consisting of a silicene layer, a graphene layer and substrates as depicted in Figure 1. The dielectric constants ϵ_1 and ϵ_2 are related to the space between the two layers and the semi-infinite region underneath the lower layer. Thus, we assume that there is no material above the upper layer whose susceptibility is $\chi_1(q,\omega)$, and it is always assumed to be a vacuum above this top layer.

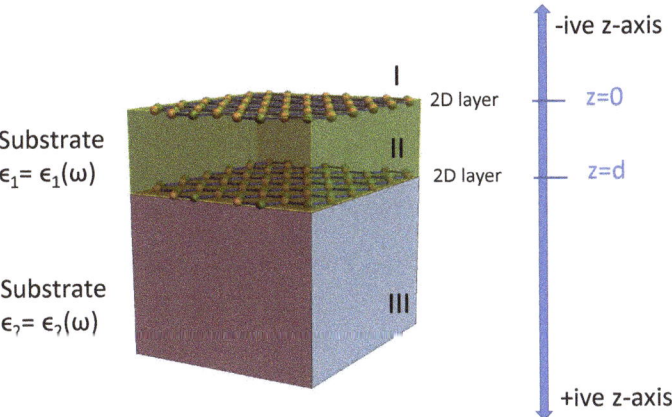

Figure 1. (Color online) Schematic illustration of a heterostructure consisting of a pair of 2D layers separated by a dielectric medium $\epsilon_1(\omega)$. This structure lies on a substrate with a dielectric function $\epsilon_2(\omega)$.

The two 2D layers may be identical or different, possessing different material properties (graphene or silicene in our case), which is reflected in their energy dispersions though the presence or absence of a band gap. The two layers could also have different or equal doping levels (Fermi energies). By employing the boundary condition of continuity of the electrostatic potential and the discontinuity of the electric field across the interface separating two media, we solved for the various coefficients appearing in the potential. Consequently, the result for the surface response function $g(q, \omega)$ gives the required conditions for the plasmon dispersion for our case, namely:

$$\begin{aligned} \phi_<(z) &= e^{-qz} - g(q,\omega)e^{qz}, z \lesssim 0, \\ \phi_>(z) &= a_1 e^{-qz} + b_1 e^{qz}, 0 \leq z \leq d, \\ \phi_{1>}(z) &= k_1 e^{-qz}, z \geq d. \end{aligned} \qquad (7)$$

Here, $\phi_<(z)$, $\phi_>(z)$ and $\phi_{1>}(z)$ correspond to the electrostatic potential of regions (I),(II) and (III), respectively, as shown in Figure 1. In order to conduct numerical computation, we make use of linear response theory, for which we have the charge density, $\sigma_1 = \chi_1 \phi_<(0)$, $\sigma_2 = \chi_2 \phi_{1>}$, with χ_1, χ_2 2D susceptibilities. Generalizing, $\chi_i = e^2 \Pi_i^0$ for convenience, we obtain the solution of these equations for different coefficients, leading to:

$$g(q,\omega) = \frac{1}{D(q,\omega)} \left\{ [q\epsilon_0(\epsilon_1 - 1) - \chi_1][q\epsilon_0(\epsilon_1 + \epsilon_2) - \chi_2] - [q\epsilon_0(\epsilon_1 + 1) + \chi_1][q\epsilon_0(\epsilon_1 - \epsilon_2) + \chi_2]e^{-2dq} \right\} \qquad (8)$$

$$D(q,\omega) \equiv [q\epsilon_0(\epsilon_1 - 1) - \chi_1][q\epsilon_0(\epsilon_1 + \epsilon_2) - \chi_2] - [q\epsilon_0(\epsilon_1 - 1) + \chi_1][q\epsilon_0(\epsilon_1 - \epsilon_2) + \chi_2]e^{-2dq}, \qquad (9)$$

where $\epsilon_1(\omega)$ is the dielectric function of the substrate between layers "1" and "2", χ_1 and χ_2 correspond to the susceptibilities of these two layers and d is the thickness of the substrate. This substrate thickness is in the order of the wavelength of light considered, for a visible light it could be of the order of a few hundred nanometers. For a thick substrate, the thickness could go up to micrometer in size, and accordingly, the plasmonic mode is modified. The plasmon dispersion equation is obtained from the solutions of $D(q, \omega) = 0$, which we solve below. We note that when we set $\chi_2 = 0$ and take the limit $d \to \infty$, Equation (8) gives the well-established form [64]:

$$g_{2D}(q,\omega) = 1 - \left\{ \frac{\epsilon_1 + 1}{2} - \frac{\chi_1}{2q\epsilon_0} \right\}^{-1}, \qquad (10)$$

which is the surface response function for a 2D layer embedded in a medium whose average background dielectric constant is $\epsilon_b = (\epsilon_1 + 1)/2$. The plasma resonances, which Equation (10) gives from its poles, are in agreement with the zeros of the dielectric function in Equation (5).

2.4. Damping Rate

We now turn to a critical issue in this paper, which concerns the rate of damping of the plasmon modes by the single-particle excitations. If this rate of damping for a plasmon mode with frequency Ω_p is denoted as γ, then $D(\Omega_p + i\gamma, q) = 0$ is the complex frequency space. Carrying out a Taylor series expansion of both the real and imaginary parts, we have:

$$\begin{aligned} D(\Omega_p + i\gamma, q) &= \text{Re } D(\Omega_p + i\gamma, q) + i\text{Im } D(\Omega_p + i\gamma, q) \\ &= \text{Re}D(\Omega_p) + i\gamma \frac{\partial}{\partial \Omega}\text{Re}D(\Omega)\Big|_{\omega=\Omega_p} + i\text{Im}D(\Omega_p) - \gamma \frac{\partial}{\partial \Omega}\text{Im}D(\Omega)\Big|_{\omega=\Omega_p} + \cdots \end{aligned} \quad (11)$$

Therefore, setting the function in Equation (11) equal to zero, we obtain γ to the lowest order as:

$$\gamma = -\frac{\text{Im}D(\Omega_p)}{\partial \text{Re}D(\omega)/\partial \omega|_{\Omega_p}}. \quad (12)$$

With these formal results, we now evaluate the plasma spectra for a double layer heterostructure. The expression shows the dependence of γ on the imaginary part of $D(\Omega_p)$ and the Real part of $D(\omega)$; which in turn, are dependent on the type of layer and the substrate considered. Eventually, we can infer that the viability of plasmon modes can be tuned by the dielectric substrate thickness and by the choice of 2D layer. In addition, the rate of decay also helps us in maintaining the intensity and the frequency of the obtained plasmon mode. This could have great impact in the development of the quantum information sharing technology and the data storing devices.

3. Numerical Results and Discussion

In our numerical calculations, energy is scaled in units of $E_F^{(0)}$ and the wave vector is scaled with $k_F^{(0)} = \sqrt{\pi n}$, which is in the experimental range for electron/hole doping densities $n = 10^{10}$ per cm^2. This gives $k_F^{(0)} = 10^6$ per cm and $E_F^{(0)}$ is equivalent to $\sim$60 meV. From the preceding discussion, in Section 2, it is clear that the plasmon modes for any system are given by the zeros of the dielectric function obtained from Equation (9). Thus, making use of this dispersion equation, we computed the plasmon mode dispersion relation for a heterostructure based on graphene and silicene with/without a substrate. For this, we first obtained a graphical result for graphene, as shown in Figure 2. One can clearly see that a single branch plasmon mode originates from the origin in $q - \omega$ space, which increases monotonically and is subject to Landau damping when the plasmon mode reaches the interband particle-hole excitation region. Figure 2 is plotted for monolayer graphene embedded in material with background dielectric constant $\epsilon_1 = \epsilon_2$, which we set equal to 1 for simplicity. The situation is similar ($\epsilon_1 = \epsilon_2 = 1$) for Figures 3–7. In Figure 2, the plasmon branches for two values of the Fermi energy are shown in panels (a,b). The damping rates of these plasmon modes are demonstrated in panel (c,d), where it is distinctly shown by an arrow pointing at the boundary of the region where Landau damping takes place. The rate of decay for both types of graphene are monotonically increasing, signifying that the deeper into the single particle excitation region plasmon mode enters, the larger the rate of decay becomes. This implies that the lifetime of the plasmon mode is decreased in the same manner.

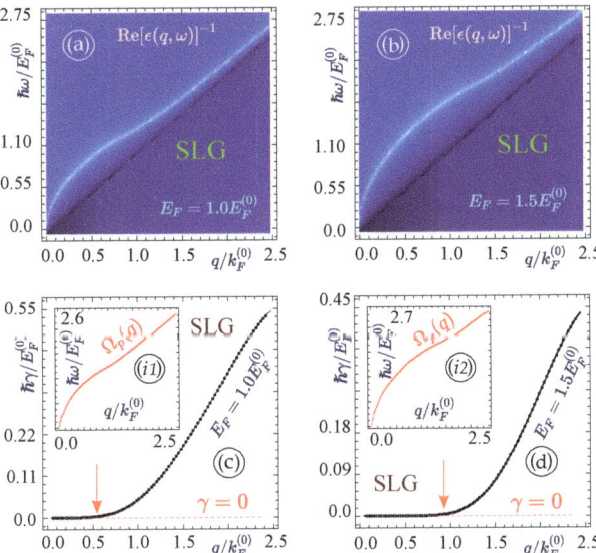

Figure 2. (Color online) Plasmon frequency $\omega_p(q)$ and damping rates $\gamma(\omega_p(q), q)$ for an isolated graphene layer (SLG) with $E_F = 1.0 E_F^{(0)}$ (left panels (**a,c**)) and $E_F = 1.5 E_F^{(0)}$ (right panels (**b,d**)). Two top panels (**a,b**) demonstrate the plasmon dispersion (either damped or undamped obtained as $Re(\epsilon(q,\omega)) = 0$, the lower plots (**c,d**) describe the corresponding damping rate along the plasmon branches, also calculated and shown as insets (i1) and (i2).

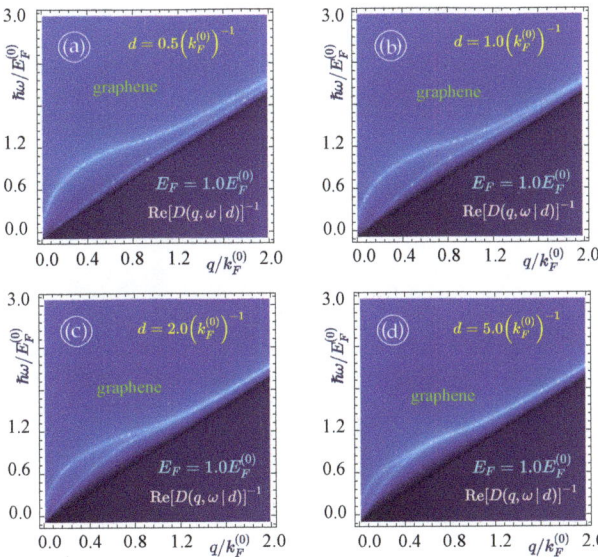

Figure 3. (Color online) Acoustic (**lower**) and optical (**upper**) plasmon modes for a pair of identical graphene layers with $E_F = 1.0 E_F^{(0)}$. Each panel corresponds to different values of the separation between the layers corresponding to (**a**) $d = 0.5(k_F^{(0)})^{-1}$, (**b**) $1.0(k_F^{(0)})^{-1}$, (**c**) $2.0(k_F^{(0)})^{-1}$ and (**d**) $5.0(k_F^{(0)})^{-1}$, as labeled. The plasmon dispersion (either damped or undamped) is obtained by solving $Re(D(q,\omega|d)) = 0$.

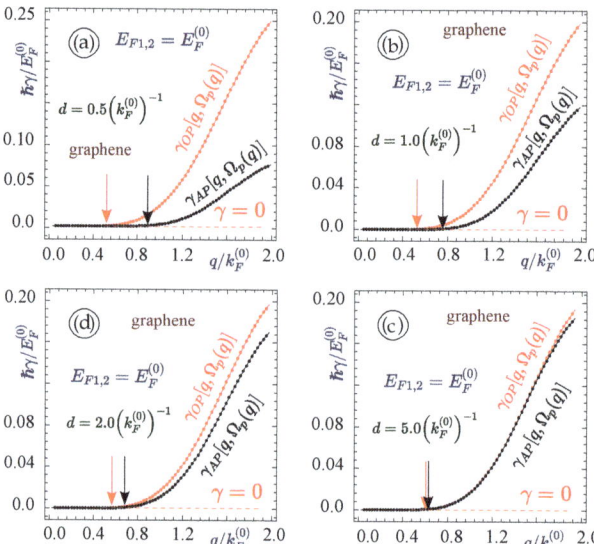

Figure 4. (Color online) The damping rates corresponding to the acoustic and optical plasmon branches shown in Figure 4 for two identical graphene layers with $E_F = 1.0 E_F^{(0)}$. Each panel corresponds to a different values of the separation between the layers with (**a**) $d = 0.5(k_F^{(0)})^{-1}$, (**b**) $1.0(k_F^{(0)})^{-1}$, (**c**) $2.0(k_F^{(0)})^{-1}$ and (**d**) $5.0(k_F^{(0)})^{-1}$, as labeled.

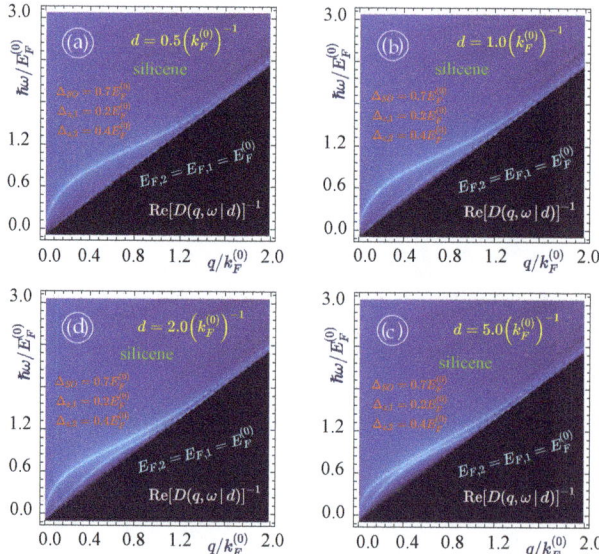

Figure 5. (Color online) Acoustic (**lower**) and optical (**upper**) plasmon dispersions for two silicene layers with $E_F = 1.0 E_F^{(0)}$ and the band gaps $\Delta_{SO,1} = \Delta_{SO,2} = 0.7 E_F^{(0)}$, $\Delta_{z,1} = 0.2 E_F^{(0)}$ and $\Delta_{z,2} = 0.4 E_F^{(0)}$. Each panel corresponds to different values of the separation between the layers (**a**) $d = 0.5(k_F^{(0)})^{-1}$, (**b**) $1.0(k_F^{(0)})^{-1}$, (**c**) $2.0(k_F^{(0)})^{-1}$ and (**d**) $5.0(k_F^{(0)})^{-1}$, as labeled.

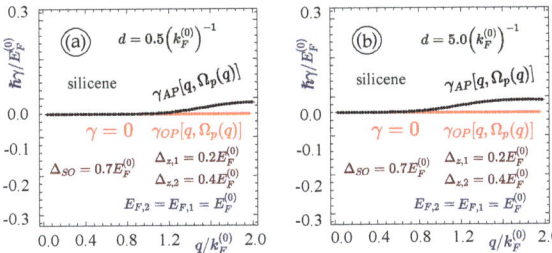

Figure 6. (Color online) The damping rates corresponding to the acoustic and optical plasmon branches calculated in Figure 6 for two silicene layers with $E_F = 1.0 E_F^{(0)}$, and the band gaps $\Delta_{SO,1} = \Delta_{SO,2} = 0.7 E_F^{(0)}$, $\Delta_{z,1} = 0.2 E_F^{(0)}$ and $\Delta_{z,2} = 0.4 E_F^{(0)}$. Each panel corresponds to different values of the separation between layers with (**a**) $d = 0.5(k_F^{(0)})^{-1}$ and (**b**) $5.0(k_F^{(0)})^{-1}$, as labeled.

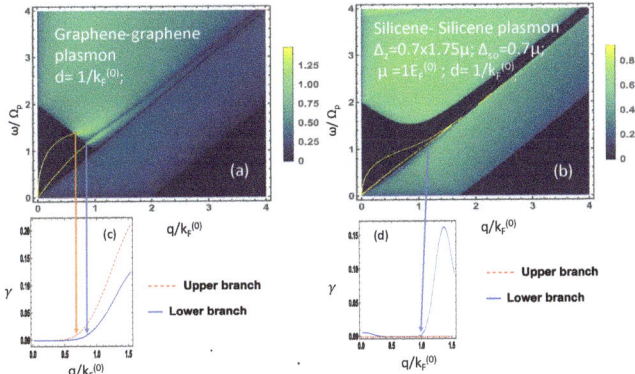

Figure 7. (Color online) Plasmon modes for (**a**) graphene–graphene structure and (**b**) silicene–silicene structure with corresponding plasmon damping rate in (**c**) and (**d**), respectively.

Going next to the case when we have a structure with two graphene layers together separated by various distances, a set of plots (as shown in Figure 3) is obtained with two branches of plasmon modes originating from the origin in the $q - \omega$ space. One can see that when the graphene sheets are brought closer, the plasmon modes move further apart, signifying weak interactions between the modes. In Figure 3a–d, plasmon modes for a structure with two graphene layers separated by a distance of $0.5(k_F^{(0)})^{-1}$, $1.0(k_F^{(0)})^{-1}$, $2.0(k_F^{(0)})^{-1}$ and $5.0(k_F^{(0)})^{-1}$ are shown, respectively, and all the plots portray that the further apart the graphene layers are, the closer the two plasmon branches become. For this same set of figures with the other parameters remaining the same, the plasmon decay rate is shown in Figure 4, which shows that the plasmon decay for the lower plasmon branch always starts at a larger wave vector value in comparison to the upper plasmon branch. As the distance of separation is increased, the two plasmon branches come closer and their decay also starts from the same value of the wave vector and the rate of decay is almost the same in value.

Now, in addition, we carried out an investigation of the plasmon modes and their decay rate for the structure with two silicene layers for various separations. The graphical results for these calculations are shown in Figure 5 where we again have two plasmon modes originating from the origin of the $q - \omega$ plane. As in the case of a two-graphene-layer structure, we again notice a similar effect on two plasmon branches coming closer to each other when their separation increases. This is demonstrated in Figure 5a–d for the layers separation distance of $0.5(k_F^{(0)})^{-1}$, $1.0(k_F^{(0)})^{-1}$, $2.0(k_F^{(0)})^{-1}$ and $5.0(k_F^{(0)})^{-1}$, respectively.

As a representative calculation, we investigated the decay rate of plasmon modes for silicene–silicene structure when their separation is $d = 0.5(k_F^{(0)})^{-1}$ and $d = 5.0(k_F^{(0)})^{-1}$. Figure 6 shows that the upper plasmon mode does not decay at all and the lower plasmon branch decays after reaching a critical wave vector. This behavior is due to the presence of a band gap for silicene, resulting in an opening in the single particle excitation region, which provides a larger area in $q - \omega$ space for the plasmon mode to survive. The upper plasmon branch in this case has a larger space and is more likely to self-sustain for a longer period without damping. On the other hand, the lower plasmon branch enters the intraband single-particle excitation region where it decays. The rate of decay starts from a critical value of the wave vector and the magnitude of this decay rate monotonically increases.

A comparison of plasmon modes and their decay for graphene–graphene and silicene–silicene structures is shown along with the single-particle excitation regions in Figure 7. The figure in panel (a) of Figure 7 shows that two plasmon modes that stem from the origin of the frequency-momentum space increase steadily and decay when it reaches the boundary of the single-particle excitation region. Corresponding red and blue lines are drawn to further clarify the point where the actual decay begins. The dark triangular region is the area where the plasmon mode survives without Landau damping and mathematically, in this region, the imaginary part of the polarization function of graphene is zero. This means that the plasmon mode has self-sustaining oscillations. The green region where the imaginary part of the polarization function is nonzero is the single particle excitation region where the plasmon mode decays into particle-hole mode. The corresponding decay rate figure, below this panel, shows that the rate of decay for the upper plasmon branch is greater and its critical wave vector is smaller compared to the lower plasmon branch.

Similar plots for silicene–silicene structures were demonstrated in Figure 7b where one can see the opening of a gap in the single-particle excitation region yielding two parts, which is a significant effect arising from the band gap. The imaginary part of the polarization function in this gap region is zero where the plasmon mode can sustain its oscillation for a long time. The upper breakaway region is a single-particle excitation region due to interband transitions of electrons from the valence to the conduction band and the lower breakaway region is the intraband single-particle excitations region, which is due to transitions within the same band from below to above the Fermi level. In Figure 7b, two plasmon modes originate from the origin as demonstrated in the figure. The upper plasmon mode survives without damping over a wide range of wave vectors and the plasmon branch enters the gap created by the opening within the single-particle excitation region. The corresponding decay rate appearing below the plasmon dispersion shows that the upper plasmon branch does not decay at all, whereas the lower plasmon branch with the part in closer contact with the intraband single-particle excitation region has a small plasmon decay rate as illustrated in the corresponding figure in the panel of Figure 7d below. As the plasmon mode rises, it is separated from the single-particle excitation region where the decay rate is zero and as it moves further away from the origin, the plasmon branch becomes closer to the single-particle excitation region where we notice the Landau damping again. Correspondingly, the decay rate increases monotonically and reaches a maximum before dropping down, indicating the reappearance of an undamped plasmon branch at a larger value of the wave vector. Another noticeable effect observed here is the closeness of the plasmon branches and the plasmon decay rate, which can be altered by altering the layer separation; this effect may be used as another plasmon mode tuning parameter.

These two branches appear as a result of the Coulomb interaction between the two layers, which couple the plasmon excitations arising on each layer. Therefore, the resulting plasmons are physically similar to two coupled oscillators. The obtained plasmon branches are defined as acoustic (lower frequencies) and optical (higher frequencies), or as in-phase and out-of-phase [65]. The number of branches is equal to the number of layers [66], or, more precisely, the number of separate plasmon excitations in them. However, one cannot attribute one branch to the first layer and the other to the second one.

We should also mention that the frequency of the optical plasmon branch depends on the distance between the layers. One can easily verify this by analyzing how far the branch moves from the main diagonal $\omega = v_F q$ when the distance between the layers is increased. However, this dependence is not as dramatic as for the acoustic plasmons, which also change their shape when the distance between the layers is increased, and the Coulomb coupling is faded away. Finally, we should say that the main subject of the present paper is the plasma branches for a system of different layers and their damping rates, which are affected by the distance between the layers and the type of substrate in between.

To extract more information about the plasmonic behavior, in Figure 8a, we have presented the figure to show the result highlighting the changes in the plasmonic nature for a structure with different types of layers and substrates. In Figure 8a, we demonstrate the plasmon mode for a structure with silicene and graphene separated by a distance of $1.0(k_F^{(U)})^{-1}$ with the vacuum in between. One can observe two plasmon modes originating from the origin in $q - \omega$ space.

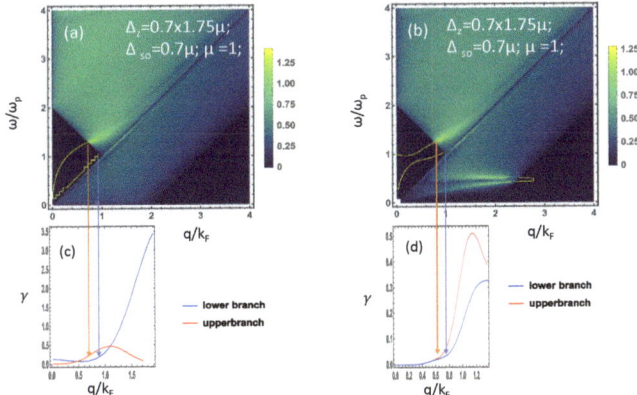

Figure 8. (Color online) Plasmon mode dispersion for a heterostructure. (**a**) Graphene-vacuum-silicene, (**b**) graphene-substrate-silicene with the dielectric function, $\epsilon_1(\omega) = 1 - \Omega_p^2/\omega^2$, for the substrate. Panels (**c**,**d**) represent the plasmon damping rate for the structure corresponding to the plasmon modes in (**a**) and (**b**), respectively. The bottom substrate is given by $\epsilon_2 = 1$ for both cases.

A special effect of overcoming the single-particle excitation region of silicene by the single-particle excitation region of graphene is observed, which causes the shortening of the plasmon branches that used to be there for the silicene–silicene structure. As soon as the plasmon branch reaches the particle-hole mode region, the plasmon mode decays by replacing one silicene layer with a graphene layer in the silicene–silicene structure. The effect, due to the band gap in silicene, is just nullified. In other words, the plasmon modes in the regime, where they used to have plasmon modes before, now do not have them, because the plasmon mode decays into the particle-hole mode of graphene. Furthermore, the result of adding a substrate between the silicene and graphene layer is illustrated in panel (b) of Figure 8. In this case, we could see a new plasmon branch originating from the bulk plasma frequency and one plasmon branch originating from the origin. Here, due to the presence of a substrate, the lower plasmon branch bends sharply toward the intraband single-particle excitation region where it decays causing complete disappearance. The upper plasmon branch and the plasmon from the bulk plasmon frequency become closer and move toward the interband single-particle excitation region where they become damped.

The results in Figure 8 correspond to a graphene layer located on the top and the silicene layer located below it in the heterostructure. The order in which the layers are placed affects the plasma dispersions only if asymmetric substrates are involved. In contrast, the plasmon dispersion relation for a pair of Coulomb-coupled layers embedded in a uniform background is determined by a 2×2 determinant equation [67], which is symmetric to switching the layers. This is what our Equation (9) is reduced to when $\epsilon_1 = \epsilon_2 = 1$ and the background dielectric constants are independent of the frequency ω.

These new effects on the plasmon branches in this type of structure were not reported previously. Results of this type help develop electronic and quantum computing devices where knowledge of the plasmonic behavior of materials and their damping nature is very essential.

4. Concluding Remarks

In summary, we have investigated the key properties of the plasmonic mode and damping for different combinations of graphene and silicene layers. The effect of the addition of substrate in between the two layers is further analyzed. This resulted in the development of a novel technique of tuning the plasmon excitation mode associated with the two dimensional systems. In our system, a complete new plasmon branch emerges from the bulk plasmon frequency, this would be very helpful in engineering modern computing devices.

Another discovery we have from this study is the disappearance of the lower plasmon branch and the suppression of the silicene band gap effect. Along with the study of interesting features in the plasmon modes, we have also developed an approach for calculating the decay rates for the plasmons due to Landau damping by the particle-hole modes.

The principal goal of our investigation and the main results of our work are concerned with the understanding of plasmonic nature and analyzing their Landau damping rates in a multi-layer structure. In contrast to Ref. [67], which includes only the long-wavelength limit for graphene without any discussion of gapped materials (such as silicene), our work is concerned with a thorough and detailed investigation of plasmon and damping rates for finite-value wave vectors and energies.

Additionally, our results infer that the number of plasmon branches emerging from the origin can be varied by choosing the number of the 2D layer. In brief, we can say that our study gives an important idea about the plasmonic behavior of a graphene–silicene-based heterostructure, which would be very helpful in carrying out further studies of other types of heterostructure, including a variety of low dimensional materials.

Author Contributions: Methodology, D.D., G.G. and A.I.; Supervision, C.-S.T.; Writing—original draft, D.D., G.G., C.-S.T. and A.I.; Writing—review & editing, D.D., G.G. and A.I. All authors have read and agreed to the published version of the manuscript.

Funding: G.G. would like to acknowledge the support from the Air Force Research Laboratory (AFRL) through Grant No. FA 9453-21-1-0046. A.I. would like to acknowledge the funding received from TRADA-52-113, PSC-CUNY Award # 64076-00 52.

Institutional Review Board Statement: Not applicable.

Informed Consent Statement: Not applicable.

Data Availability Statement: Exclude this statement as the study did not report any data.

Conflicts of Interest: The authors declare no conflict of interest.

References

1. Vogt, P.; Padova, P.D.; Quaresima, C.; Avila, J.; Frantzeskakis, E.; Asensio, M.C.; Resta, A.; Ealet, B.; Lay, G.L. Silicene: Compelling experimental evidence for graphenelike two-dimensional silicon. *Phys. Rev. Lett.* **2012**, *108*, n155501. [CrossRef] [PubMed]
2. Takeda, K.; Shiraishi, K. Theoretical possibility of stage corrugation in Si and Ge analogs of graphite. *Phys. Rev. B* **1994**, *50*, 14916. [CrossRef] [PubMed]
3. Novoselov, K.S.; Geim, A.K.; Morozov, S.V.; Jiang, D.; Zhang, Y.; Dubonos, S.V.; Grigorieva, I.V.; Firsov, A.A. Electric field effect in atomically thin carbon films. *Science* **2004**, *306*, 666. [CrossRef] [PubMed]
4. Ni, Z.; Liu, Q.; Tang, K.; Zheng, J.; Zhou, J.; Qin, R.; Gao, Z.; Yu, D.; Lu, J. Tunable bandgap in silicene and germanene. *Nano Lett.* **2012**, *12*, 113. [CrossRef]

5. Drummond, N.D.; Zolyomi, V.; Fal'Ko, V.I. Electrically tunable band gap in silicene. *Phys. Rev. B* **2012**, *85*, 075423. [CrossRef]
6. Liu, J.; Zhang, W. Bilayer silicene with an electrically-tunable wide band gap. *RSC Adv.* **2013**, *3*, 21943. [CrossRef]
7. Kane, C.L.; Mele, E.J. Quantum spin Hall effect in graphene. *Phys. Rev. Lett.* **2005**, *95*, 226801. [CrossRef]
8. Yao, Y.; Ye, F.; Qi, X.L.; Zhang, S.C.; Fang, Z. Spin-orbit gap of graphene: First-principles calculations. *Phys. Rev. B* **2007**, *75*, 041401. [CrossRef]
9. Bao, W.S.; Liu, S.Y.; Lei, X.L.; Wang, C.M. Nonlinear dc transport in graphene. *J. Phys. Condens. Matter* **2009**, *21*, 305302. [CrossRef]
10. Dora, B.; Moessner, R. Nonlinear electric transport in graphene: Quantum quench dynamics and the Schwinger mechanism. *Phys. Rev. B* **2010**, *81*, 165431. [CrossRef]
11. Tian, S.; Wang, P.; Liu, X.; Zhu, J.; Fu, H.; Taniguchi, T.; Watanabe, K.; Chen, J.H.; Lin, X. Nonlinear transport of graphene in the quantum Hall regime. *2D Mater.* **2016**, *4*, 015003. [CrossRef]
12. Rosenstein, B.; Lewkowicz, M.; Kao, H.C.; Korniyenko, Y. Ballistic transport in graphene beyond linear response. *Phys. Rev. B* **2010**, *81*, 041416. [CrossRef]
13. Dahal, D.; Gumbs, G. Effect of energy band gap in graphene on negative refraction through the veselago lens and electron conductance. *Phys. Chem. Solids* **2017**, *100*, 83–91. [CrossRef]
14. Wakamura, T.; Gueron, S.; Bouchiat, H. Novel transport phenomena in graphene induced by strong spin-orbit interaction. *arXiv* **2021**, arXiv:2112.07813.
15. Kim, W.Y.; Kim, K.S. Carbon nanotube, graphene, nanowire, and molecule-based electron and spin transport phenomena using the nonequilibrium Green's function method at the level of first principles theory. *J. Comp. Chem.* **2008**, *29*, 1073–1083. [CrossRef] [PubMed]
16. Yan, X.Z.; Romiah, Y.; Ting, C.S. Electric transport theory of Dirac fermions in graphene. *Phys. Rev. B* **2008**, *77*, 125409. [CrossRef]
17. Gumbs, G.; Iurov, A.; Horing, N.J.M. Non-local plasma spectrum of graphene interacting with a thick conductor. *Phys. Rev. B* **2015**, *91*, 235416. [CrossRef]
18. Gumbs, G.; Balassis, A.; Dahal, D.; Glasser, M.L. Thermal smearing and screening in a strong magnetic field for Dirac materials in comparison with the two dimensional electron liquid. *Eur. Phys. J. B* **2016**, *89*, 234. [CrossRef]
19. Checkelsky, J.G.; Ong, N.P. Thermopower and Nernst effect in graphene in a magnetic field. *Phys. Rev. B* **2009**, *80*, 081413. [CrossRef]
20. Nakamura, M. Orbital magnetism and transport phenomena in two-dimensional Dirac fermions in a weak magnetic field. *Phys. Rev. B* **2007**, *76*, 113301. [CrossRef]
21. Yan, J.; Zhang, Y.; Kim, P.; Pinczuk, A. Electric field effect tuning of electron-phonon coupling in graphene. *Phys. Rev. Lett.* **2007**, *98*, 166802. [CrossRef] [PubMed]
22. Yu, Y.J.; Zhao, Y.; Ryu, S.; Brus, L.E.; Kim, K.S.; Kim, P. Tuning the graphene work function by electric field effect. *Nano Lett.* **2009**, *9*, 3430. [CrossRef] [PubMed]
23. Liu, X.; Li, Z. Electric field and strain effect on graphene-MoS2 hybrid structure: Ab initio calculations. *J. Phys. Chem. Lett.* **2015**, *6*, 3269. [CrossRef]
24. Eletskii, A.V.; Iskandarova, I.M.; Knizhnik, A.A. Graphene fabrication methods and thermophysical properties. *Physics-Uspekhi* **2011**, *54*, 227. [CrossRef]
25. Jia, X.; Campos-Delgado, J.; Terrones, M.; Meunier, V.; Dresselhaus, M.S. Graphene edges: A review of their fabrication and characterization. *Nanoscale* **2011**, *3*, 86. [CrossRef]
26. Cai, J.; Ruffieux, P.; Jaafar, R.; Bieri, M.; Braun, T.; Blankenburg, S.; Muoth, M.; Seitsonen, A.P.; Saleh, M.; Feng, X.; et al. Atomically precise bottom-up fabrication of graphene nanoribbons. *Nature* **2010**, *466*, 470. [CrossRef]
27. Aliofkhazraei, M.; Ali, N.; Milne, W.I.; Ozkan, C.S.; Mitura, S.; Gervasoni, J.L.; (Eds.) *Graphene Science Handbook: Fabrication Methods*; CRC Press: Boca Raton, FL, USA, 2016.
28. Dadkhah, N.; Vazifehshenas, T.; Farmanbar, M.; Salavati-Fard, T. A theoretical study of collective plasmonic excitations in double-layer silicene at finite temperature. *J. App. Phys.* **2019**, *125*, 104302. [CrossRef]
29. Men, N.V. Plasmon modes in N-layer silicene structures. *J. Phys. Cond. Matt.* **2021**, *34*, 8. [CrossRef]
30. De Abajo, F.J.G. Graphene plasmonics: Challenges and opportunities. *ACS Phot.* **2014**, *1*, 135. [CrossRef]
31. Farmer, D.B.; Rodrigo, D.; Low, T.; Avouris, P. Plasmon-plasmon hybridization and bandwidth enhancement in nanostructured graphene. *Nano Lett.* **2015**, *15*, 2582–2587. [CrossRef]
32. Gumbs, G.; Horing, N.J.; Iurov, A.; Dahal, D. Plasmon excitations for encapsulated graphene. *J. Phys. App. Phys.* **2016**, *49*, 225101. [CrossRef]
33. Iurov, A.; Zhemchuzhna, L.; Dahal, D.; Gumbs, G.; Huang, D. Quantum-statistical theory for laser-tuned transport and optical conductivities of dressed electrons in α-T_3 materials. *Phys. Rev. B* **2020**, *101*, 035129. [CrossRef]
34. Iurov, A.; Gumbs, G.; Huang, D.; Silkin, V.M. Plasmon dissipation in gapped graphene open systems at finite temperature. *Phys. Rev. B* **2016**, *93*, 035404. [CrossRef]
35. Iurov, A.; Gumbs, G.; Huang, D.; Zhemchuzhna, L. Controlling plasmon modes and damping in buckled two-dimensional material open systems. *J. Appl. Phys.* **2017**, *121*, 084306. [CrossRef]
36. Gumbs, G.; Iurov, A.; Wu, J.Y.; Lin, M.F.; Fekete, P. Plasmon excitations of multi-layer graphene on a conducting substrate. *Sci. Rep.* **2016**, *6*, 21063. [CrossRef] [PubMed]

37. Dahal, D.; Gumbs, G.; Huang, D. Effect of strain on plasmons, screening, and energy loss in graphene/substrate contacts. *Phys. Rev. B* **2018**, *98*, 045427. [CrossRef]
38. Iurov, A.; Zhemchuzhna, L.; Gumbs, G.; Huang, D.; Fekete, P.; Anwar, F.; Dahal, D.; Weekes, N. Tailoring plasmon excitations in $\alpha - \mathcal{T}_3$ armchair nanoribbons. *Sci. Rep.* **2021**, *11*, 20577. [CrossRef]
39. Raza, A.; Ikram, M.; Aqeel, M.; Imran, M.; Ul-Hamid, A.; Riaz, K.N.; Ali, S. Enhanced industrial dye degradation using Co doped in chemically exfoliated MoS2 nanosheets. *Appl. Nanosci.* **2020**, *10*, 1535–1544. [CrossRef]
40. Ikram, M.; Ali, S.; Aqeel, M.; Ul-Hamid, A.; Imran, M.; Haider, J.; Haider, A.; Shahbaz, A.; Ali, S.X. Reduced graphene oxide nanosheets doped by Cu with highly efficient visible light photocatalytic behavior. *J. Alloy. Compd.* **2020**, *837*, 155588. [CrossRef]
41. Dong-Thi, K.P.; Nguyen, V.M. Plasmonic Excitations in 4-MLG Structures: Background Dielectric Inhomogeneity Effects. *J. Low Temp. Phys.* **2022**, *206*, 51–62. [CrossRef]
42. Calafell, I.A.; Cox, J.D.; Radonji, M.; Saavedra, J.R.M.; de Abajo, F.G.; Rozema, L.A.; Walther, P. Quantum computing with graphene plasmons. *NPJ Quant. Inf.* **2019**, *5*, 1–7. [CrossRef]
43. Hanson, G.W.; Gangaraj, S.H.; Lee, C.; Angelakis, D.G.; Tame, M. Quantum plasmonic excitation in graphene and loss-insensitive propagation. *Phys. Rev. A* **2015**, *92*, 013828. [CrossRef]
44. Christensen, T.; Wang, W.; Jauho, A.P.; Wubs, M.; Mortensen, N.A. Classical and quantum plasmonics in graphene nanodisks: Role of edge states. *Phys. Rev. B* **2014**, *90*, 241414. [CrossRef]
45. Esfandiari, M.; Jarchi, S.; Nasiri-Shehni, P.; Ghaffari-Miab, M. Enhancing the sensitivity of a transmissive graphene-based plasmonic biosensor. *App. Opt.* **2021**, *60*, 1201–1208. [CrossRef]
46. Tong, J.; Jiang, L.; Chen, H.; Wang, Y.; Yong, K.T.; Forsberg, E.; He, S. Graphene-bimetal plasmonic platform for ultra-sensitive biosensing. *Opt. Commun.* **2018**, *410*, 817–823. [CrossRef]
47. Hu, W.; Huang, Y.; Chen, C.; Liu, Y.; Guo, T.; Guan, B.O. Highly sensitive detection of dopamine using a graphene functionalized plasmonic fiber-optic sensor with aptamer conformational amplification. *Sens. Actuat. Chem.* **2018**, *264*, 440–447. [CrossRef]
48. Andoy, N.M.; Filipiak, M.S.; Vetter, D.; Sanz, O.G.; Tarasov, A. Graphene-Based Electronic Immunosensor with Femtomolar Detection Limit in Whole Serum. *Adv. Mat. Technol.* **2018**, *3*, 1800186. [CrossRef]
49. Viswanathan, S.; Narayanan, T.N.; Aran, K.; Fink, K.D.; Paredes, J.; Ajayan, P.M.; Renugopalakrishanan, V. Graphene-protein field effect biosensors: Glucose sensing. *Mat. Today* **2015**, *18*, 513–522. [CrossRef]
50. Huang, A.; Li, W.; Shi, S.; Yao, T. Quantitative fluorescence quenching on antibody-conjugated graphene oxide as a platform for protein sensing. *Sci. Rep.* **2017**, *7*, 40772. [CrossRef]
51. Goncalves, S.P.; Peres, N. *I An introduction to Graphene Plasmonics*; World Scientific: Singapore, 2016.
52. Gumbs, G.; Dahal, D.; Balassis, A. Effect of Temperature and Doping on Plasmon Excitations for an Encapsulated Double-Layer Graphene Heterostructure. *Phys. Stat. Solidi (b)* **2018**, *255*, 1700342. [CrossRef]
53. Hwang, E.H.; Sarma, S.D. Plasmon modes of spatially separated double-layer graphene. *Phys. Rev. B* **2009**, *80*, 205405. [CrossRef]
54. Ezawa, M. A topological insulator and helical zero mode in silicene under an inhomogeneous electric field. *New J. Phys.* **2012**, *14*, 033003. [CrossRef]
55. Tabert, C.J.; Nicol, E.J. Valley-spin polarization in the magneto-optical response of silicene and other similar 2D crystals. *Phys. Rev. Lett.* **2013**, *110*, 197402. [CrossRef] [PubMed]
56. Neto, A.C.; Guinea, F.; Peres, N.M.; Novoselov, K.S.; Geim, A.K. The electronic properties of graphene. *Rev. Mod. Phys.* **2009**, *81*, 109. [CrossRef]
57. Wunsch, B.; Stauber, T.; Sols, F.; Guinea, F. Dynamical polarization of graphene at finite doping. *New J. Phys.* **2006**, *8*, 318. [CrossRef]
58. Balassis, A.; Dahal, D.; Gumbs, G.; Iurov, A.; Huang, D.; Roslyak, O. Magnetoplasmons for the $\alpha - T_3$ model with filled Landau levels. *J. Phys. Cond. Matt.* **2020**, *32*, 485301. [CrossRef] [PubMed]
59. Roldan, R.; Goerbig, M.O.; Fuchs, J.N. The magnetic field particle-hole excitation spectrum in doped graphene and in a standard two-dimensional electron gas. *Semicond. Sci. Technol.* **2010**, *25*, 034005. [CrossRef]
60. Patel, D.K.; Ashraf, S.S.; Sharma, A.C. Finite temperature dynamical polarization and plasmons in gapped graphene. *Phys. Stat. Solidi (b)* **2015**, *252*, 1817–1826. [CrossRef]
61. Ramezanali, M.R.; Vazifeh, M.M.; Asgari, R.; Polini, M.; MacDonald, A.H. Finite-temperature screening and the specific heat of doped graphene sheets. *J. Phys. Math. Theor.* **2009**, *42*, 214015. [CrossRef]
62. Tabert, C.J.; Nicol, E.J. Dynamical polarization function, plasmons, and screening in silicene and other buckled honeycomb lattices. *Phys. Rev. B* **2014**, *89*, 195410. [CrossRef]
63. Pyatkovskiy, P.K. Dynamical polarization, screening, and plasmons in gapped graphene. *J. Phys. Cond. Matt.* **2008**, *21*, 025506. [CrossRef] [PubMed]
64. Persson, B.N.J. Inelastic electron scattering from thin metal films. *Solid State Commun.* **1984**, *52*, 811–813. [CrossRef]
65. Hwang, H.E.; Sarma, S.D. Dielectric function, screening, and plasmons in two-dimensional graphene. *Phys. Rev. B* **2007**, *75*, 205418. [CrossRef]
66. Zhu, J.J.; Badalyan, S.M.; Peeters, F.M. Plasmonic excitations in Coulomb-coupled N-layer graphene structures. *Phys. Rev. B* **2013**, *87*, 085401. [CrossRef]
67. Sarma, S.D.; Li, Q. Intrinsic plasmons in two-dimensional Dirac materials. *Phys. Rev. B* **2013**, *87*, 235418. [CrossRef]

Review

Review on Graphene-, Graphene Oxide-, Reduced Graphene Oxide-Based Flexible Composites: From Fabrication to Applications

Aamir Razaq [1,*], Faiza Bibi [1], Xiaoxiao Zheng [2], Raffaello Papadakis [3,4], Syed Hassan Mujtaba Jafri [5] and Hu Li [2,6,*]

1. Department of Physics, COMSATS University Islamabad, Lahore Campus 54000, Pakistan; faiza4563@gmail.com
2. Shandong Technology Centre of Nanodevices and Integration, School of Microelectronics, Shandong University, Jinan 250101, China; 202120353@mail.sdu.edu.cn
3. TdB Labs AB, Uppsala Business Park, 75450 Uppsala, Sweden; rafpapadakis@gmail.com
4. Department of Chemistry, Uppsala University, 75120 Uppsala, Sweden
5. Department of Electrical Engineering, Mirpur University of Science and Technology (MUST), Mirpur Azad Jammu and Kashmir 10250, Pakistan; hassan.jafri@must.edu.pk
6. Ångström Laboratory, Department of Materials Science and Engineering, Uppsala University, 75121 Uppsala, Sweden
* Correspondence: Aamirrazaq@cuilahore.edu.pk (A.R.); Hu.Li@sdu.edu.cn (H.L.)

Abstract: In the new era of modern flexible and bendable technology, graphene-based materials have attracted great attention. The excellent electrical, mechanical, and optical properties of graphene as well as the ease of functionalization of its derivates have enabled graphene to become an attractive candidate for the construction of flexible devices. This paper provides a comprehensive review about the most recent progress in the synthesis and applications of graphene-based composites. Composite materials based on graphene, graphene oxide (GO), and reduced graphene oxide (rGO), as well as conducting polymers, metal matrices, carbon–carbon matrices, and natural fibers have potential application in energy-harvesting systems, clean-energy storage devices, and wearable and portable electronics owing to their superior mechanical strength, conductivity, and extraordinary thermal stability. Additionally, the difficulties and challenges in the current development of graphene are summarized and indicated. This review provides a comprehensive and useful database for further innovation of graphene-based composite materials.

Keywords: graphene; flexible devices; composite; graphene oxide; reduced graphene oxide

Citation: Razaq, A.; Bibi, F.; Zheng, X.; Papadakis, R.; Jafri, S.H.M.; Li, H. Review on Graphene-, Graphene Oxide-, Reduced Graphene Oxide-Based Flexible Composites: From Fabrication to Applications. *Materials* **2022**, *15*, 1012. https://doi.org/10.3390/ma15031012

Academic Editors: Victoria Samanidou and Eleni Deliyanni

Received: 6 January 2022
Accepted: 25 January 2022
Published: 28 January 2022

Copyright: © 2022 by the authors. Licensee MDPI, Basel, Switzerland. This article is an open access article distributed under the terms and conditions of the Creative Commons Attribution (CC BY) license (https://creativecommons.org/licenses/by/4.0/).

1. Introduction

It is well known that materials play an important role in the development of science and technology, because the realization of a new technology often requires the support of novel materials. Therefore, exploring materials with excellent properties has always been an important subject of scientific research. A remarkable material, graphene, has attracted widespread attention since it was first exfoliated from graphite by Andre Geim and Konstantin Novoselov in 2004. As a result of its prominent performances, graphene can be used in various fields, such as energy storage, biosensing, optoelectronics, flexible electronics, electrochemical sensing, robotics, textile industry, and so on [1–4]. The discovery of graphene marked the beginning of a new era in material science research [5].

Graphene with a thickness of a single carbon atom is arranged in a honeycomb lattice. It is very solid and can be fashioned into 0D, 1D, 3D forms (Figure 1) [6]. In addition, it is extraordinarily transparent and possesses high crystallite as well as outstanding electronic properties. Although graphene has many excellent properties, there is no bandgap in graphene, and it has poor water solubility, which greatly limits its application in some areas [7]. An effective way to overcome these limitations and expand the range of application of graphene is to prepare graphene derivatives. For example, treating graphite with strong oxidants will add epoxy groups, hydroxyl groups, and carboxyl group on the basal plan of graphite layers, thus producing graphene oxide (GO). These polar oxygen-containing functional groups make GO highly hydrophilic. This allows GO to have excellent dispersibility in many solvents, especially in water. In addition, the oxygen-containing functional groups can provide reactive sites for chemical modification or functionalization of GO, which in turn can be used to develop GO-based materials. Although the oxygen-containing groups can obviate some disadvantages of graphene, they also cause some problems. For example, they make GO electrically insulating. Nevertheless, the chemical reduction of GO can restore its conductivity to some extent. The obtained reduced GO (rGO) still carries some functional groups, which results in a good dispersion of rGO in many solvents. Most importantly, it is relatively easy to control the electrical performance and solubility of rGO by controlling the number of the remaining functional groups. The properties of this chemically reduced graphene approximately resemble those of pristine graphene [6,8]. The transformation of graphite to graphite oxide, GO and graphene is shown in Figure 2.

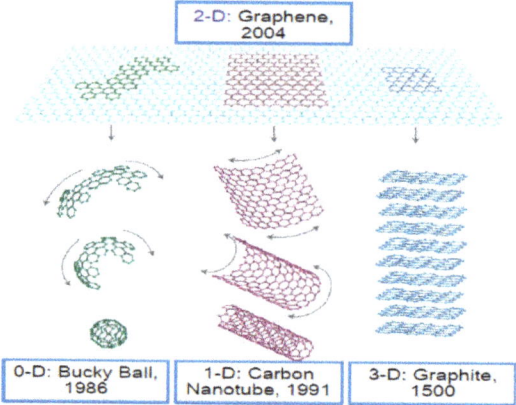

Figure 1. Various structures of graphene (0D bucky ball, 1D carbon nanotube, 3D graphite). (Reproduced with permission from ref. [9]. Copyright 2016 Springer Publications).

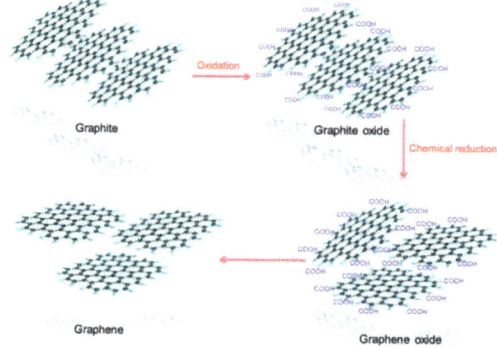

Figure 2. Schematic of the transformation of graphite oxide to GO and graphene. (Reproduced with permission from ref. [6]. Copyright 2016 SAGE Publications).

Graphene and its derivatives have their own unique advantages and can be used in many domains applying different techniques, such as thermal chemical vapor deposition (CVD), self-assembly technique, spin coating, vacuum filtration, thermal decomposition, solution dispersion technique, and chemical decomposition polymer processing technique [8,10–15]. This review comprises of synthesis of composite material based on graphene and its derivatives along with chemical properties, and mainly focus on the potential applications of these materials.

2. Synthesis of Graphene and Its Derivatives

2.1. GO

There are many reports about the synthesis of GO, and the structures of the obtained products are slightly different (Figure 3). One of the most classical methods was proposed by Williams Hummers JR and Richard Offeman in 1958. The general process was as follows. Graphite was first mixed with concentrated sulfuric acid and oxidizers such as sodium nitrate, then potassium permanganate was added under a precise temperature control, followed by the addition of reducing and reaction stopping agents such as hydrogen peroxide at the end of the process [16]. This method supplies a high yield of colloidal suspension and powdery product [17]. Later, numerous research groups made further improvements of the preparation method focusing on three main parameters, i.e., precursors ratio, time, and temperature [18]. For example, Marcano et al. synthesized GO by using the Tours method and obtained high-quality GO by adding phosphoric acid as a key precursor and removing sodium nitrate with the product. This method is better than previous methods due to its simplicity and outstanding product quality [19]. In addition to these three parameters, the size of graphite particle also has a great effect on the quality of the final products [20]. According to the demands of different applications, various physical forms of GO such as suspension, powder, and flexible sheet can be prepared. The corresponding photos are shown in Figure 4.

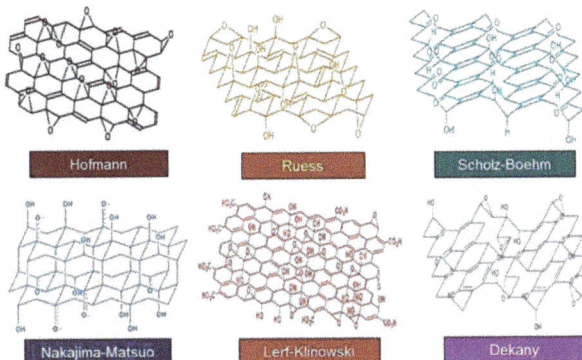

Figure 3. Different structures of GO. (Reproduced with permission from ref. [6]. Copyright 2016 SAGE Publications).

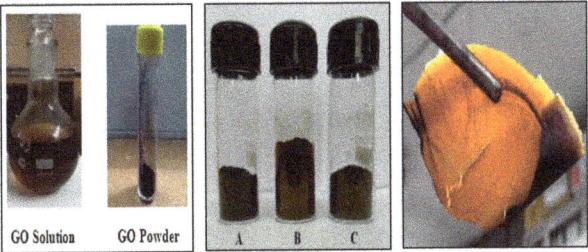

Figure 4. GO suspension (left side), powder, and flexible sheet (right side). (Reproduced with permission from ref. [18,21]. Copyright 2015 JNMNT and 2014 Hindawi Publications).

2.2. Graphene

Several attempts have been adopted to study the synthesis methods of carbon-based materials. The very first attempt dates back to 1962. Boehm et al. prepared soot composed of thin-layer graphite-intercalated compounds by the reduction and combustion of graphite oxide. In 1944, these products were named as graphene platelets and had a single carbon layer [22]. In 2004, Novoselov et al. obtained graphene by the scotch tape method and won the Noble Prize in 2010 [23]. Till now, the preparation methods of graphene include top-down and bottom-up techniques. Top-down methods include scotch tape exfoliation, liquid-phase exfoliation, and chemical synthesis. Bottom-up methods mainly comprise CVD and molecular beam epitaxy [24–26]. In this section, typical methods of graphene synthesis will be introduced.

2.2.1. Exfoliation and Cleavage

Micromechanical cleavage is a process in which the bonds in graphite crystal are broken by mechanical energy so that graphene sheets are peeled from a silicon substrate. Exfoliation can be done in solution by intercalating graphite and exfoliating to a single carbon sheet [27]. GO prepared by the conventional Hummer's method acts as a precursor for the preparation of graphene sheets when intercalated with sulphuric acid. The process involves the reduction and expansion of sulphuric acid-intercalated graphite oxide for the large-scale production of graphene. As shown in Figure 5, when the slurry obtained by Hummer's method is placed in a box furnace, graphite oxide can be expanded into graphene [28].

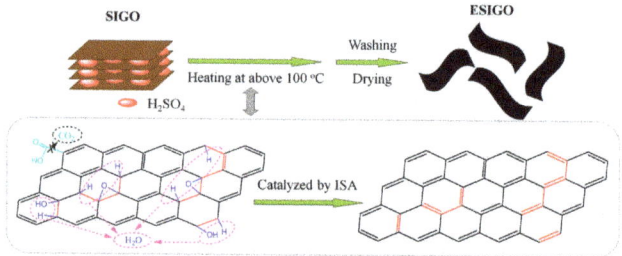

Figure 5. Schematic of the H_2SO_4-intercalated GO (SIGO) process. (Reproduced with permission from ref. [28]. Copyright 2013 Springer Nature Publications).

2.2.2. Thermal CVD Techniques

The CVD technique is another common method used to synthesize graphene. Certain carrier gases and carbon-based precursors like camphor and methane are injected into a CVD chamber at a specific temperature. Then, the carbon precursor is decomposed to form graphene on transition metal sheets such as a nickel foam (Figure 6). In addition to exfoliation and the CVD method, there are many other methods that can be used to prepare graphene, such as thermal decomposition of SiC [10] and others.

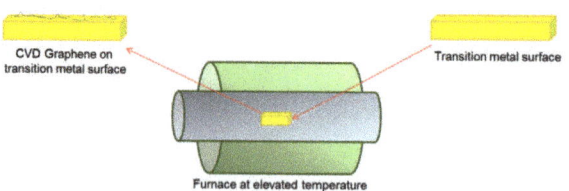

Figure 6. Graphene synthesis via the CVD method. (Reproduced with permission from ref. [6]. Copyright 2016 SAGE Publications).

2.3. rGO

GO prepared by Hummer's method consists of a-few-layer carbon platelets decorated with oxygen containing functional groups. The removal of some oxygen-based groups by reducing agents or thermal treatment can yield rGO (Figure 7). The main process is as follows. GO is exfoliated via ultrasonication and then reduced by hydrazine hydrate, a strong reducing agent, for 2 h. Since hydrazine is toxic, alternative reagents such as $NaBH_4$, ascorbic acid, and HI can be used. Among these, ascorbic acid is essential for the scalable production of rGO. The chemical procedure to obtain rGO using ascorbic acid as a reducing agent is shown in Figure 8. This reaction does not produce toxic gases [29]. rGO has been proven to be a good candidate for various applications such as field effect transistors (FET), solar cells, energy applications, and production of composite paper-like materials [30] due to its abundant atomic defects.

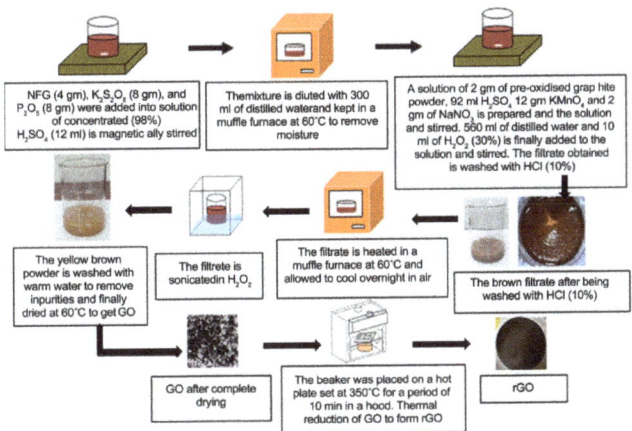

Figure 7. Steps of the synthesis of GO and rGO. (Reproduced with permission from ref. [29]. Copyright 2017 Scientific Research Publications).

Figure 8. Consecutive steps in the chemical synthesis of rGO using ascorbic acid as a reducing agent. (**a**) Oxidation and exfoliation of graphite using Hummer's method. (**b**) Reduction and conversion of Mn (VII) ions to soluble Mn (II) ions by the addition of ascorbic acid. (**c**) Color transition of the exfoliated graphite oxide from greenish yellow to black in the early stage of reduction. (**d**) Loss of hydrophilicity of GO when stirring is paused. (**e**) Precipitation of rGO after completion of the reduction stage and cooling down to room temperature. (**f**) Filtration of rGO using cellulose filter paper. (**g**) rGO powder after freeze-drying. (Reproduced with permission from ref. [31]. Copyright 2015 Springer Nature Publications).

3. Flexible Graphene Composites: From Fabrication to Applications

With the evolution of science and technology, more and more novel materials with fascinating properties have been discovered and can be applied in many domains. Among these novel materials, graphene has received a lot of attention because of its excellent properties, such as high mechanical strength, stability, charge storage capacity, etc. Furthermore, graphene has very good flexibility and shows excellent application prospects in some flexible composite materials. For instance, a flexible composite consisting of polyethylene–ioxythiophene–graphene was fabricated by the following method. First, $PtCl_4$ was added to an NaOH solution under stirring followed by heating at 160 °C for 3 h. Next, the solution was treated with 2 M sulphuric acid and ethyl glycol and then was electrochemically deposited on a graphene-filtrated carbon cloth/graphene paper substrate. It is worth noting that this flexible composite material is expected to be used in energy storage, because the square shape of the corresponding electrochemical graph indicates excellent capacitive properties [32]. Likewise, when polyaniline (PANI), a conducting polymer with good stability, was mixed with graphene in the form of nanofibers by the vacuum filtration method, the obtained composite film showed not only excellent flexibility but also good electrochemical stability [33].

Many additional related reports on the preparation and application of other flexible graphene-based composites have been published [34]. For example, by regulating the ratio of each components, poly(3,4-ethylenedioxythiophene)/poly(4-styrenesulfonate) (PEDOT/PSS)/graphene composites can be fabricated; they show great potential applications in energy-harvesting systems such as thermoelectric devices and solar cells [35]. Besides, a flexible composite was prepared by simple coating MnO_2 on Zn_2SnO_4 (ZTO) nanowires grown on carbon microfibers. This material can be used in supercapacitor electrodes, whose composite analysis suggests a long cycle life [36]. A typical rectangular voltammogram can be seen for carbon cloth and graphene-coated carbon cloth with electrodeposited PEDOT. This result indicates that graphene-based materials have excellent electrical performance and can be excellent electrode materials in energy storage devices. A simple spin coating technique used at ambient conditions for the fabrication of graphene-based transparent electrodes was proposed. In this method, a graphene slurry was added to dimethyl sulfoxide (DMSO) and then to a pure PEDOT/PSS aqueous solution. Then, a spin coater was used to spin the coating, and the product was left to rest at room temperature [37]. Graphene/MnO_2 combined with light-weight carbon nanotubes (CNTs) formed an ultra-flexible thin-film composite, which has been used for various energy storage devices as a robust electrode, as it holds extraordinary mechanical properties with superb electrochemical activities when fabricated by the chemical co-precipitation method [38].

Beside flexibility, the light weight and the efficiency of a device are also very important. To meet the current energy demand and increase the performance of energy devices, paper-based electrodes of graphene/PANI composite have been reported. They were prepared by the electropolymerization of PANI on graphene paper [39]. As shown in Figure 9a,b, graphene/PANI paper retains the origin flexibility of graphene paper. Graphene/PANI paper as a supercapacitor electrode exbibits a high specific capacitance and excellent cycling stability due to the uniform growth of PANI on graphene (Figure 9c–f); it has great potential for application in the construction of portable energy devices. Light-weight and flexible graphene/polypyrrole (PPy) fibers were fabricated by spinning GO and pyrrole in a $FeCl_3$ solution, which helped to control the diameter of fiber, finally obtaining graphene/PPy fibers [40]. MnO_2 can also be used for the fabrication of this composite due to its high specific capacitance. A 3D graphene/MnO_2 composite foam to be used as a negative electrode for asymmetric supercapacitors was fabricated by the solution casting method. GO was reduced on Ni foam and then subjected to electrodeposition of MnO_2 to obtain an asymmetric supercapacitor, showing excellent cyclic stability (Figure 10) [41].

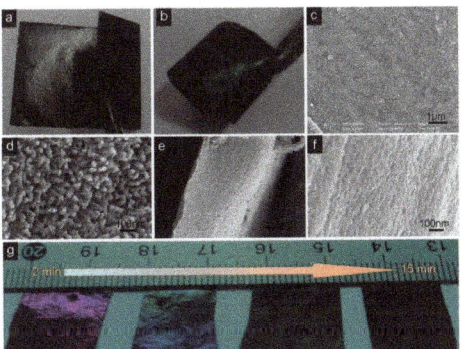

Figure 9. (**a**) Flexible graphene paper with the size of 8 × 5 cm. (**b**) Graphene/PANI paper (3 cm × 1.5 cm), electrochemical deposition time of 10 min. (**c**,**d**) SEM images of the surface of graphene/PANI paper at different magnifications. (**e**,**f**) SEM images of cross sections of graphene/PANI paper at different magnifications. (**g**) Graphene/PANI composite papers with different electropolymerization times (From left to right: 2, 5, 10, 15 min). (Reproduced with permission from ref. [39]. Copyright 2013 RSC Publications).

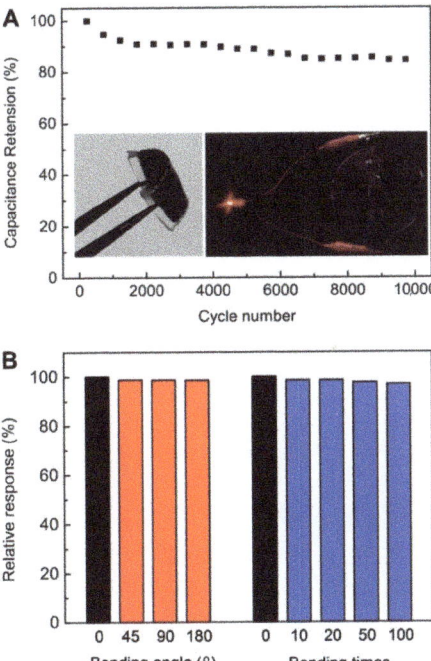

Figure 10. (**A**) Cyclic behavior of MnO_2/ERGO//CNT ERGO. (**B**) Specific capacitance retention ratio of the flexible supercapacitor after inward bending by different angles or repeated bending. (Reproduced with permission from ref. [41]. Copyright 2014 Wiley Publications).

The incorporation of graphene-based composites provides an innovative way for wearable electronics and energy storage devices. Various techniques have been used for the synthesis of these composites. For example, a hydrothermal approach can be applied to fabricate a textile-base graphene composite as an electrode. First, graphene is transferred onto a polyester fabric, and then the graphene/polyester/MnO_2 composite is placed in an autoclave at 140 °C. Finally, the product is washed with deionized water and dried in an oven. The composite reveals good electrochemical performance with high mechanical stability [42]. TiO_2 is also another promising electro-active metal oxide. For instance, it was used to fabricate the material for a supercapacitor electrode. The fabrication process of the TiO_2/graphene/PPy composite for energy applications is as follows. At different temperatures, titian as a starting precursor, was mixed with chemically modified graphene. After drying, electrodeposition of PPy was carried out. As illustrated in Figure 11, the composite revealed increased capacitance and cycling stability [43].

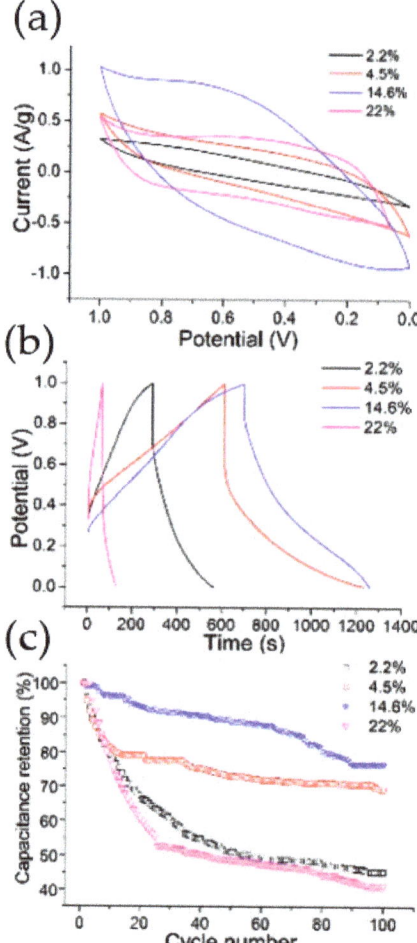

Figure 11. Electrochemical performances of TiO_2/graphene/PPy with different TiO_2 content: (**a**) CV curves; (**b**) galvanostatic charge–discharge curves; (**c**) cycle stability. (Reproduced with permission from ref. [43]. Copyright 2015 ACS Publications).

Multiple graphene-based composites including epoxy/graphene, polystyrene/graphene, polyaniline/graphene, nafion/graphene, poly(3,4-ethyldioxythiophene)/graphene, polyethylene terephthalate/graphene, and polycarbonate/graphene nanocomposites have been fabricated through in situ intercalative polymerization, solution intercalation, as well as melt intercalation [44,45]. In addition, a flexible graphene/MnO_2 composite for paper electrodes was prepared by three steps, during which the GO/MnO_2 composite was obtained by dispersion. Composite paper was obtained by vacuum filtration followed by thermal reduction [46]. Beside physical synthesis routes, CVD is also a good approach for the fabrication of materials. Therefore, a graphene composite with porous carbon was fabricated via CVD on a Ni gauze substrate, which had excellent compatibility because of the porosity of the composite [47]. Likewise, the hydrothermal method is commonly used on account of its simplicity. For example, $ZnFe_2O_4$ nanoparticles treated with nitrogen-doped reduced graphene were reported as suitable in energy application, specifically for supercapacitors [48]. To attain maximum charge storage and long cycle durability, another composite of 3D graphene/NiOOH/Ni_3S_2 was fabricated in two steps. First, 3D graphene was prepared on the surface of nickel foam by the CVD method. Second, the composite was generated by the hydrothermal method [49].

Although the material choice for certain application remains crucial, the choice of the substrate has a great effect on flexibility. Textile fibers, carbon cloth, and paper pulp have evolved as excellent substrate choices for various graphene-based composites. For example, a graphene-based carbon cloth composite fabricated by the simple brush coating technique showed great properties as an electrode material [50]. Light weight, ultrathin, and flexible electrodes with outstanding mechanical and electrochemical properties are needed of today. As shown in Figure 12, a cellulose fiber-based graphene paper composite was obtained by the dipping and drying method via the hydrothermal route, and possesses environment-friendly and cost-effective features [51].

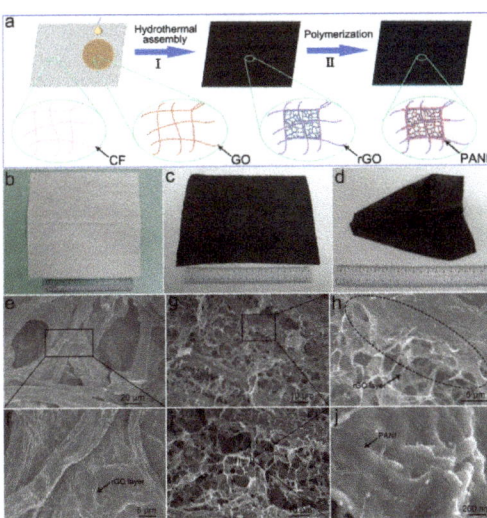

Figure 12. (a) Schematic diagram of the preparation of PANI-rGO/cellulose fiber composite paper. Optical images of (b) pure cellulose fiber paper and (c,d) nanostructured rGO/cellulose fiber composite paper. SEM images of (e,f) rGO-coated cellulose fiber paper, (g,h) nanostructured rGO/cellulose fiber composite paper, and (i,j) PANI-rGO/cellulose fiber composite paper. (Reproduced with permission from ref. [51]. Copyright 2014 Wiley Publications).

The composite of vanadium oxide with graphene paper is binder-free and shows versatility. The fabrication follows an alkaline deoxygenation process, which is more suitable than the chemical reduction of GO to graphene. This flexible composite paper membrane possesses remarkable advantages for double-layered and pseudocapacitive electrodes [52]. In order to avoid toxicity effects, dimethylformamide was utilized instead of hydrazine for the reduction of GO, obtaining outstanding efficiency along with flexibility. Graphene nanosheets were combined with carbon nanofibers to form a composite with enhanced properties via electrospinning, which is favorable for energy applications [53]. In addition to the capacitive properties of carbon-based materials, the mechanical and thermal properties of flexible composite materials obtained from conductive graphene/poly(vinyl chloride) (graphene/PVC) films have also been studied. PVC and graphene sheets were mixed together by liquid dispersion and dripped onto cells, followed by drying in an oven. The acquired composite possesses good thermal stability [54]. Additionally, the properties of conductive polymers like PANI are remarkably enhanced by the addition of graphene-based composites. The formation of a graphene/polyaniline flexible composite can be obtained via in situ anodic electro polymerization. Graphene paper (Figure 13) was directly used as a working electrode in PANI electrolyte, washed, and dried after the complete process, recording a high capacitance [55].

Figure 13. Flexible graphene paper. (Reproduced with permission from ref. [55]. Copyright 2009 ACS Publications.)

Moreover, combing graphene with some common substances in nature can lead to composites with outstanding performances. As shown in Figure 14, lignocellulose/graphene conductive paper composite, which worked as a good active electrode, was fabricated through a simple and time-efficient technique by the one-pot method [56].

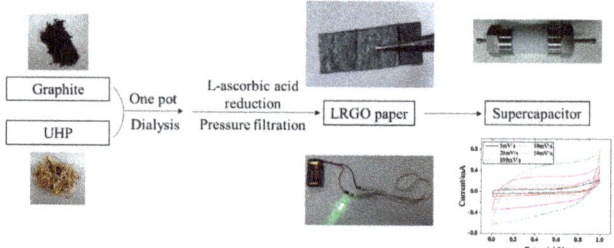

Figure 14. Supercapacitor derived from a conductive paper consisting of a lignocellulose/rGO (LRGO) composite. (Reproduced with permission from ref. [56]. Copyright 2018 Springer Nature Publications).

A nano nickel oxide/graphene PANI composite with enhanced cyclic stability and a high specific capacitance of 92% after 2500 charge–discharge cycles can be applied in the fabrication of energy storage devices [57]. Consequently, graphene-based polymer/metal oxide composite paper electrodes show enhanced electrochemical performance and great potential for application in portable electronics industry.

4. Flexible Composites and Applications of GO/rGO

The designed flexible composites should have not only excellent flexibility but also a certain strength which would enable the composites to withstand external environmental factors. Recently, a layered composite of GO/PVC with large mechanical strength fabricated by the vacuum filtration method was reported [58]. These flexible composites composed of GO and rGO are useful in multiple applications including energy storage, water purification, textiles, and robotics. Polymerization was used to prepare PANI nanowires on a GO sheet composite. The obtained products showed excellent performance when used as supercapacitor electrodes [59,60]. Water purification has become another issue in the past few decades. Ongoing research has tackled this problem. For example, a GO-based TiO_2 composite membrane could be used as a filtration membrane for the removal of water impurities. The composite was fabricated by vacuum filtration and allowed a moderate water purification [61]. Although chemically modified graphene or rGO itself is not very appealing in terms of its properties, these properties can be enhanced in forming composite materials with conductive polymers. For instance, an rGO/polypyrrole nanowires composite fabricated in situ showed better performances than rGO and can be used in the fabrication of portable electronic devices [62]. Yarns were used to produce an electronic textile fabric by coating rGO through electrostatic self-assembly in the presence of adhesive bovine serum albumin. The preparation of the material is shown in Figure 15 [63].

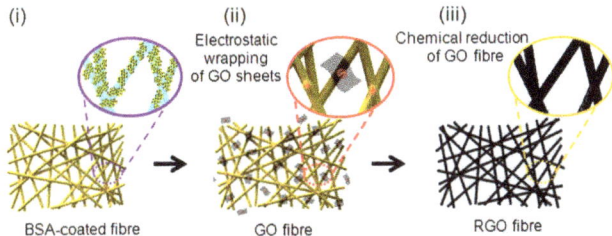

Figure 15. Illustration of the three steps used to prepare rGO/nano yarns (rGO/NYs). (Reproduced with permission from ref. [63]. Copyright 2013 Wiley Publications).

E-textile has revolutionized the whole flexible and portable device industry with the introduction of additional properties. In addition, a pressed composite of rGO–$MnFe_2O_4$ and polyvinylidene fluoride fabricated by a simple sonication method turned out to be a good absorber of harmful microwaves in the electromagnetic spectrum [64]. Within energy applications, flexible composites of V_2O_5/polyindole and activated carbon cloth were used as cathodic and anodic electrodes of an asymmetric supercapacitor. V_2O_5 nanostructures were constructed on a carbon cloth by in situ polymerization and showed good cyclic stability on testing (Figure 16) [65].

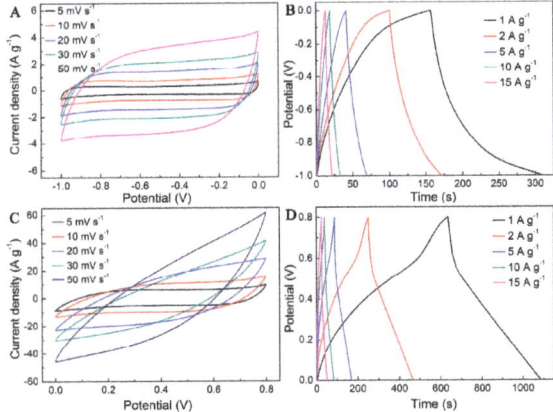

Figure 16. CVs of (**A**) rGO@actived carbon cloth (rGO@ACC) and (**C**) V_2O_5/polyindole@ACC. Galvanostatic charge–discharge curves of (**B**) rGO@ACC and (**D**) V_2O_5/polyindole@ACC. (Reproduced with permission from ref. [65]. Copyright 2016 ACS Publications).

Paper electrodes for energy application are receiving great attention. A GO solution was prepared by Hummer's method. Then, GO-based paper electrodes which can act as flexible substrates, actuators, supercapacitor electrodes etc., were fabricated through the steps of vacuum infiltration, spin coating, and drop casting.

A composite of nickel cobalt oxide/GO was tested as a supercapacitor electrode and revealed a large capacitance of 1211 Fg^{-1}. Its fabrication was achieved by coprecipitation using sodium dodecyl sulfate as the template and ammonia as the precipitant [66]. rGO obtained from GO by the hydrothermal route and titanium carbide obtained by selective etching of aluminum were combined by ultra-sonication and filtration, yielding the rGO/titanium carbide composite. CV, GCD, and EIS analysis proved it to be outstanding for electrochemical performance in supercapacitors [67]. Additionally, other carbon-based materials like CNT are extraordinary products due to several characteristics when hybridized with conducting polymers such as MnO_x and rGO composites developed by spray coating and electrodeposition. They show a high capacitive behavior and improve the cyclic stability for supercapacitors [68]. Notably, graphene and its derivatives are analogous to other carbon-based materials, providing new perspectives to research. The fabrication techniques are also being modified. Recently, the metal-organic framework template-assisted method was utilized on a large scale and showed great potential for energy applications. In this regard, as shown in Figure 17, rGO/MoO_3 was reported to be an excellent composite for energy storage in supercapacitors as an electrode [69].

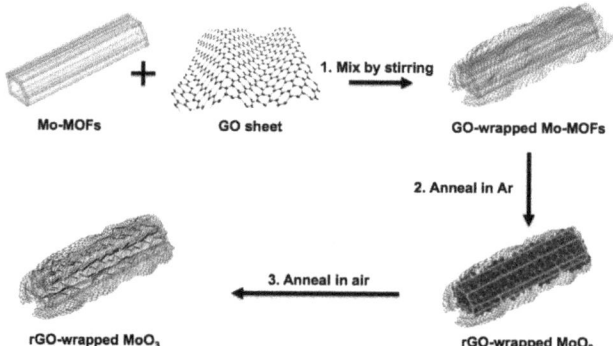

Figure 17. Preparation process of rGO/MoO_3 composites. (Reproduced with permission from ref. [69]. Copyright 2015 Wiley Publications).

Organic substrates have been widely used in recent studies, but metallic substrates remain an important choice. Copper metallic foil acted as a substrate for the growth of rGO/Cu_2O through the hydrothermal technique and showed moderate advantages as an electrode in supercapacitors [70]. However, the sol–gel approach is also simple and has been used for the fabrication of rGO-based composites. rGO paper was obtained by a modified Hummer's method followed by evaporation drying. ZnO was deposited in the form of layers on rGO paper by using a stabilizer through a synthesis process. The composite ZnO/rGO/ZnO has been utilized for supercapacitor electrodes [71]. The choice of the material for positive or negative electrodes plays a vital role in energy devices. A compatible negative electrode material in a supercapacitor for Fe_2O_3 nanoparticle clusters/rGO paper was investigated. The composite was synthesized through the hydrothermal technique [72]. Chong et al. [73] prepared an MnO_2/rGO nanocomposite by a facial one-step electrochemical method. MnO_2 nanoparticles were uniformly distributed on rGO nanosheets and acted as spacers to prevent rGO nanosheets from restacking. This unique structure provided MnO_2/rGO with high specific capacitance. Furthermore, the MnO_2/rGO composite also showed high conductivity and excellent potential cycling stability, and has potential as electrode material for highly stable supercapacitors. In addition to MnO_2, tungsten oxide (WO_x) is widely studied as electrode material for supercapacitors. Recently, a $W_{18}O_{49}$ nanowires (NWs)/rGO nanocomposite, which can act as the negative electrode in asymmetric supercapacitor devices, was prepared from the precursors WCl_6 and GO by the solvothermal method [74]. The asymmetric supercapacitor $W_{18}O_{49}$ NWs-rGO//rGO showed high specific capacitance and excellent cycling stability. For the fabrication of paper-based electrode, the incorporation of celluloses and pulps is desirable to attain flexibility and stability. GO-based nanocomposite of nanocrystalline cellulose acetate fabricated via stirring and a solvent casting method showed high thermal stability and good mechanical strength [75]. Likewise, cotton pulp was mixed with LiCl in addition with anhydrous DMAc by stirring. Further addition of a GO suspension to cotton pulp resulted in the formation of a cellulose-based composite useful for energy and memory storage [76]. An outstanding anodic electrode material was designed by fabricating a composite of GO and TiO_2, whereas further reduction of the composite to rGO/TiO_2 was obtained by stirring and drying. Anatase TiO_2 exhibits higher power and energy density than other conventional metal oxides [77]. In comparison with cellulose, the residual paper pulp is more stable. As a consequence, it can be used in the fabrication of rGO-based flexible composites. First, the paper pulp was stirred in stable solvent and then it was mixed with GO. Next, the suspension was infiltrated with and reduced by hydrazine vapors at a certain temperature via the drop casting technique. The obtained composite possessed better performance compared with cellulous-based composites when applied in flexible electrodes [78]. Altogether, natural fiber-based GO/rGO paper composites have been proven to have excellent performance in multiple applications, especially in energy storage and conversion devices in the modern portable device industry.

5. Conclusions and Perspective

With the fast development of portable, wearable, and lightweight electronic devices, highly efficient and flexible energy strategies are urgently needed. In order to achieve this goal, it is crucial to explore novel materials. Graphene has attracted tremendous attention in the field of material science due to its outstanding properties since it was first exfoliated in 2004. This review aimed to outline the different fabrication methods and applications of graphene-based materials, especially for flexible, portable, environment-friendly, and cost-effective energy storage and conversion devices. Some representative methods used to prepare the composites based on graphene and the corresponding applications are listed in Table 1. The outstanding performances of these composites are due to the special structure and excellent properties of graphene, as well as the ease of functionalization of GO and rGO. Graphene-based materials show great application potential in flexible devices. In addition, they further promote the miniaturization and portability of devices and have a huge effect on human life. For example, integrating graphene-based energy storage into wearable devices is promising for human health monitoring. Graphene-based composite membranes also show potential applications in water purification and can be used to remove dyes molecules in water. The progress in flexible composites based on graphene and its derivatives is rapid, and some achievements have been made in recent years. However, to realize graphene's practical applications, there are still many challenges to solve. For example, the large-scale production of graphene with high quality and uniform structure is still a big challenge. There are many techniques that can be used to prepare graphene, such as exfoliation from graphite and CVD techniques, but they are cumbersome, time-consuming, and expensive. Furthermore, the complicate transfer process further limits the wide application of graphene. Chemical oxidation of graphite is the most widely used method way to prepare graphene derivatives. However, the synthesis and purification procedures of the oxidation of graphite are complex and risky. It is hard to precisely control the compositions and sizes of graphene sheets, which heavily affects the performance of the composites. In addition, in order to synthesize composites with excellent performances, interfacial interactions between graphene or its derivatives and other functional materials need to be systematically studied. Although difficulties and challenges still exist, with the development of science and technology, more and more technically feasible strategies will be explored. We believe that flexible graphene-based devices and systems will emerge as essential instruments in our daily lives.

Table 1. Preparation and application of graphene-based composites.

Carbon-Based Material	Composites	Preparation Methods	Applications
graphene	poly ethylenedioxythiophene-graphene	electrochemically deposition	energy storage devices [32]
	(PEDOT:PSS)/graphene	in situ polymerization	energy harvesting systems [35]
	graphene/MnO_2/CNTs	chemical co-precipitation method	energy storage devices as a robust electrode [38]
GO/rGO	graphene/PANI	electropolymerization	paper electrode [39]
	graphene/MnO_2	electrodeposition	asymmetric supercapacitor [41]
	TiO_2/graphene/PPy	electrodeposition	supercapacitor [43].
	PANI/GO	polymerization	supercapacitor electrodes [59,60]
	GO based TiO_2 composite membrane	vacuum filtration	water purification system [61]
	nickel cobalt oxide/GO	coprecipitation	supercaps electrode [66]
	rGO/polypyrrole nanowires composite	in situ route	portable electronic devices [62]
	rGO/Cu_2O	hydrothermal technique	supercaps [70]
	paper pulp/rGO	drop casting technique	flexible electrode [78]

Funding: This research was funded by Shandong Natural Science Foundation of China (Grant No.: ZR2021QE148), Guangdong Natural Science Foundation of China (Grant No.: 2214050001485), Olle Engkvist (Grant No.: 211-0068), Swedish Research Council Formas (2019-01538), and Qilu Young Scholar Program of Shandong University (Grant No.: 11500082063141).

Institutional Review Board Statement: Not applicable.

Informed Consent Statement: Not applicable.

Data Availability Statement: No new data were created or analyzed in this study. Data sharing is not applicable to this article.

Acknowledgments: The authors would like to thank Aimin Song (Shandong University) for inspiration, showing the way, and providing substantive support.

Conflicts of Interest: The authors declare no conflict of interest.

References

1. Jalili, R.; Aboutalebi, S.H.; Esrafilzadeh, D.; Shepherd, R.L.; Chen, J.; Aminorroaya–Yamini, S.; Konstantinov, K.; Minett, A.I.; Razal, J.M.; Wallace, G.G. Scalable one-step wet-spinning of graphene fibers and yarns from liquid crystalline dispersions of graphene oxide: Towards multifunctional textiles. *Adv. Funct. Mater.* **2013**, *23*, 5345–5354. [CrossRef]
2. Lundstedt, A.; Papadakis, R.; Li, H.; Han, Y.; Jorner, K.; Bergman, J.; Leifer, K.; Grennberg, H.; Ottosson, H. White–Light Photoassisted Covalent Functionalization of Graphene Using 2–Propanol. *Small Methods* **2017**, *1*, 1700214. [CrossRef]
3. Li, H.; Papadakis, R.; Jafri, S.H.M.; Thersleff, T.; Michler, J.; Ottosson, H.; Leifer, K. Superior adhesion of graphene nanoscrolls. *Commun. Phys.* **2018**, *1*, 44. [CrossRef]
4. Huang, W.; Xia, X.; Zhu, C.; Steichen, P.; Quan, W.; Mao, W.; Yang, J.; Chu, L.; Li, X. Memristive Artificial Synapses for Neuromorphic Computing. *Nanomicro. Lett.* **2021**, *13*, 85. [CrossRef]
5. Chabot, V.; Higgins, D.; Yu, A.; Xiao, X.; Chen, Z.; Zhang, J. A review of graphene and graphene oxide sponge: Material synthesis and applications to energy and the environment. *Energy Environ. Sci.* **2014**, *7*, 1564–1596. [CrossRef]
6. Khan, Z.U.; Kausar, A.; Ullah, H.; Badshah, A.; Khan, W.U. A review of graphene oxide, graphene buckypaper, and polymer/graphene composites: Properties and fabrication techniques. *J. Plast. Film Sheeting* **2016**, *32*, 336–379. [CrossRef]
7. Geim, A.; Novoselov, K. *The Rise of Graphene, Nanoscience and Technology: A Collection of Reviews from Nature Journals*; World Scientific: Singapore, 2009; pp. 11–19.
8. Eda, G.; Chhowalla, M. Chemically derived graphene oxide: Towards large-area thin-film electronics and optoelectronics. *Adv. Mater.* **2010**, *22*, 2392–2415. [CrossRef]
9. Sajibul, M.; Bhuyan, A.; Uddin, M.N.; Islam, M.M.; Bipasha, F.A.; Hossain, S.S. Synthesis of graphene. *Int. Nano Lett.* **2016**, *6*, 65–83.

10. Choi, W.; Lahiri, I.; Seelaboyina, R.; Kang, Y.S. Synthesis of graphene and its applications: A review. *Crit. Rev. Solid State Mater. Sci.* **2010**, *35*, 52–71. [CrossRef]
11. Tsou, C.-H.; An, Q.-F.; Lo, S.-C.; De Guzman, M.; Hung, W.-S.; Hu, C.-C.; Lee, K.-R.; Lai, J.-Y. Effect of microstructure of graphene oxide fabricated through different self-ssembly techniques on 1-butanol dehydration. *J. Membr. Sci.* **2015**, *477*, 93–100. [CrossRef]
12. Naficy, S.; Jalili, R.; Aboutalebi, S.H.; Gorkin III, R.A.; Konstantinov, K.; Innis, P.C.; Spinks, G.M.; Poulin, P.; Wallace, G.G. Graphene oxide dispersions: Tuning rheology to enable fabrication. *Mater. Horiz.* **2014**, *1*, 326–331. [CrossRef]
13. Liu, Y.; Feng, J. An attempt towards fabricating reduced graphene oxide composites with traditional polymer processing techniques by adding chemical reduction agents. *Compos. Sci. Technol.* **2017**, *140*, 16–22. [CrossRef]
14. Liu, J.; Chen, S.; Papadakis, R.; Li, H. Nanoresolution patterning of hydrogenated graphene by electron beam induced C-H dissociation. *Nanotechnology* **2018**, *29*, 415304. [CrossRef]
15. Liu, J.; Papadakis, R.; Li, H. Experimental observation of size-dependent behavior in surface energy of gold nanoparticles through atomic force microscope. *Appl. Phys. Lett.* **2018**, *113*, 083108. [CrossRef]
16. Hummers, W.S., Jr.; Offeman, R.E. Preparation of graphitic oxide. *J. Am. Chem. Soc.* **1958**, *80*, 1339. [CrossRef]
17. Dikin, D.A.; Stankovich, S.; Zimney, E.J.; Piner, R.D.; Dommett, G.H.; Evmenenko, G.; Nguyen, S.T.; Ruoff, R.S. Preparation and characterization of graphene oxide paper. *Nature* **2007**, *448*, 457–460. [CrossRef]
18. Paulchamy, B.; Arthi, G.; Lignesh, B. A simple approach to stepwise synthesis of graphene oxide nanomaterial. *J. Nanomed. Nanotechnol.* **2015**, *6*, 1.
19. Marcano, D.C.; Kosynkin, D.V.; Berlin, J.M.; Sinitskii, A.; Sun, Z.; Slesarev, I.; Alemany, L.B.; Lu, W.; Tour, J.M. Improved synthesis of graphene oxide. *ACS Nano* **2010**, *4*, 4806–4814. [CrossRef]
20. Chowdhury, D.R.; Singh, C.; Paul, A. Role of graphite precursor and sodium nitrate in graphite oxide synthesis. *RSC Adv.* **2014**, *4*, 15138–15145. [CrossRef]
21. Song, J.; Wang, X.; Chang, C.-T. Preparation and characterization of graphene oxide. *J. Nanomater.* **2014**, *2014*, 276143. [CrossRef]
22. Boehm, H.; Clauss, A.; Fischer, G.O.; Hofmann, U. In Surface Properties of Extremely Thin Graphite Lamellae. In Proceedings of the Fifth Conference on Carbon, University Park, PA, USA, 19–23 June 1961; pp. 73–80.
23. McNaught, A.D.; Wilkinson, A. *Compendium of Chemical Terminology. IUPAC Recommendations*; Blackwell Science: Hoboken, NJ, USA, 1997.
24. Kumar, V.; Kumar, A.; Lee, D.J.; Park, S.S. Estimation of Number of Graphene Layers Using Different Methods: A Focused Review. *Materials* **2021**, *14*, 4590. [CrossRef]
25. Kamedulski, P.; Ilnicka, A.; Lukaszewicz, J.P. Selected Aspects of Graphene Exfoliation as an Introductory Step Towards 3D Structuring of Graphene Nano-Sheets. *Curr. Graphene Sci.* **2019**, *2*, 106–117. [CrossRef]
26. Prekodravac, J.R.; Kepić, D.P.; Colmenares, J.C.; Giannakoudakis, D.A.; Jovanović, S.P. A comprehensive review on selected graphene synthesis methods: From electrochemical exfoliation through rapid thermal annealing towards biomass pyrolysis. *J. Mater. Chem. C* **2021**, *9*, 6722–6748. [CrossRef]
27. Novoselov, K.S.; Jiang, Z.; Zhang, Y.; Morozov, S.; Stormer, H.L.; Zeitler, U.; Maan, J.; Boebinger, G.; Kim, P.; Geim, A.K. Room-temperature quantum Hall effect in graphene. *Science* **2007**, *315*, 1379. [CrossRef]
28. Hong, Y.; Wang, Z.; Jin, X. Sulfuric acid intercalated graphite oxide for graphene preparation. *Sci. Rep.* **2013**, *3*, 3439. [CrossRef]
29. Alam, S.N.; Sharma, N.; Kumar, L. Synthesis of graphene oxide (GO) by modified hummers method and its thermal reduction to obtain reduced graphene oxide (rGO). *Graphene* **2017**, *6*, 1–18. [CrossRef]
30. Ray, S. *Applications of Graphene and Graphene-Oxide Based Nanomaterials*; William Andrew: Norwich, NY, USA, 2015.
31. Abdolhosseinzadeh, S.; Asgharzadeh, H.; Kim, H.S. Fast and fully–scalable synthesis of reduced graphene oxide. *Sci. Rep.* **2015**, *5*, 10160. [CrossRef]
32. Chu, C.-Y.; Tsai, J.-T.; Sun, C.-L. Synthesis of PEDOT-modified graphene composite materials as flexible electrodes for energy storage and conversion applications. *Int. J. Hydrogen Energy* **2012**, *37*, 13880–13886. [CrossRef]
33. Wu, Q.; Xu, Y.; Yao, Z.; Liu, A.; Shi, G. Supercapacitors based on flexible graphene/polyaniline nanofiber composite films. *ACS Nano* **2010**, *4*, 1963–1970. [CrossRef]
34. Tang, P.; Han, L.; Zhang, L. Facile synthesis of graphite/PEDOT/MnO_2 composites on commercial supercapacitor separator membranes as flexible and high–performance supercapacitor electrodes. *ACS Appl. Mater. Interfaces* **2014**, *6*, 10506–10515. [CrossRef]
35. Yoo, D.; Kim, J.; Kim, J.H. Direct synthesis of highly conductive poly (3,4-ethylenedioxythiophene): Poly (4-styrenesulfonate) (PEDOT: PSS)/graphene composites and their applications in energy harvesting systems. *Nano Res.* **2014**, *7*, 717–730. [CrossRef]
36. Bao, L.; Zang, J.; Li, X. Flexible Zn_2SnO_4/MnO_2 core/shell nanocable-carbon microfiber hybrid composites for high-performance supercapacitor electrodes. *Nano Lett.* **2011**, *11*, 1215–1220. [CrossRef]
37. Chang, H.; Wang, G.; Yang, A.; Tao, X.; Liu, X.; Shen, Y.; Zheng, Z. A transparent, flexible, low-temperature, and solution-processible graphene composite electrode. *Adv. Funct. Mater.* **2010**, *20*, 2893–2902. [CrossRef]
38. Cheng, Y.; Lu, S.; Zhang, H.; Varanasi, C.V.; Liu, J. Synergistic effects from graphene and carbon nanotubes enable flexible and robust electrodes for high-performance supercapacitors. *Nano Lett.* **2012**, *12*, 4206–4211. [CrossRef]
39. Cong, H.-P.; Ren, X.-C.; Wang, P.; Yu, S.-H. Flexible graphene-polyaniline composite paper for high-performance supercapacitor. *Energy Environ. Sci.* **2013**, *6*, 1185–1191. [CrossRef]

40. Ding, X.; Zhao, Y.; Hu, C.; Hu, Y.; Dong, Z.; Chen, N.; Zhang, Z.; Qu, L. Spinning fabrication of graphene/polypyrrole composite fibers for all-solid-state, flexible fibriform supercapacitors. *J. Mater. Chem. A* **2014**, *2*, 12355–12360. [CrossRef]
41. Zhang, Z.; Xiao, F.; Qian, L.; Xiao, J.; Wang, S.; Liu, Y. Facile Synthesis of 3D MnO$_2$-Graphene and Carbon Nanotube-Graphene Composite Networks for High-Performance, Flexible, All-Solid-State Asymmetric Supercapacitors. *Adv. Energy Mater.* **2014**, *4*, 1400064. [CrossRef]
42. Guo, M.-X.; Bian, S.-W.; Shao, F.; Liu, S.; Peng, Y.-H. Hydrothermal synthesis and electrochemical performance of MnO$_2$/graphene/polyester composite electrode materials for flexible supercapacitors. *Electrochim. Acta* **2016**, *209*, 486–497. [CrossRef]
43. Jiang, L.-L.; Lu, X.; Xie, C.-M.; Wan, G.-J.; Zhang, H.-P.; Youhong, T. Flexible, free-standing TiO$_2$-graphene-polypyrrole composite films as electrodes for supercapacitors. *J. Phys. Chem. C* **2015**, *119*, 3903–3910. [CrossRef]
44. Kuilla, T.; Bhadra, S.; Yao, D.; Kim, N.H.; Bose, S.; Lee, J.H. Recent advances in graphene based polymer composites. *Prog. Polym. Sci.* **2010**, *35*, 1350–1375. [CrossRef]
45. Sun, F.; Li, H.; Leifer, K.; Gamstedt, E.K. Rate effects on localized shear deformation during nanosectioning of an amorphous thermoplastic polymer. *Int. J. Solids Struct.* **2017**, *129*, 40–48. [CrossRef]
46. Li, Z.; Mi, Y.; Liu, X.; Liu, S.; Yang, S.; Wang, J. Flexible graphene/MnO$_2$ composite papers for supercapacitor electrodes. *J. Mater. Chem.* **2011**, *21*, 14706–14711. [CrossRef]
47. Li, X.; Zang, X.; Li, Z.; Li, X.; Li, P.; Sun, P.; Lee, X.; Zhang, R.; Huang, Z.; Wang, K. Large-area flexible core-shell graphene/porous carbon woven fabric films for fiber supercapacitor electrodes. *Adv. Funct. Mater.* **2013**, *23*, 4862–4869. [CrossRef]
48. Li, L.; Bi, H.; Gai, S.; He, F.; Gao, P.; Dai, Y.; Zhang, X.; Yang, D.; Zhang, M.; Yang, P. Uniformly dispersed ZnFe$_2$O$_4$ nanoparticles on nitrogen-modified graphene for high-performance supercapacitor as electrode. *Sci. Rep.* **2017**, *7*, 43116. [CrossRef]
49. Lin, T.-W.; Dai, C.-S.; Hung, K.-C. High energy density asymmetric supercapacitor based on NiOOH/Ni$_3$S$_2$/3D graphene and Fe$_3$O$_4$/graphene composite electrodes. *Sci. Rep.* **2014**, *4*, 7274. [CrossRef]
50. Liu, W.-W.; Yan, X.-B.; Lang, J.-W.; Peng, C.; Xue, Q.-J. Flexible and conductive nanocomposite electrode based on graphene sheets and cotton cloth for supercapacitor. *J. Mater. Chem.* **2012**, *22*, 17245–17253. [CrossRef]
51. Liu, L.; Niu, Z.; Zhang, L.; Zhou, W.; Chen, X.; Xie, S. Nanostructured graphene composite papers for highly flexible and foldable supercapacitors. *Adv. Mater.* **2014**, *26*, 4855–4862. [CrossRef]
52. Perera, S.D.; Liyanage, A.D.; Nijem, N.; Ferraris, J.P.; Chabal, Y.J.; Balkus, K.J., Jr. Vanadium oxide nanowire-Graphene binder free nanocomposite paper electrodes for supercapacitors: A facile green approach. *J. Power Sources* **2013**, *230*, 130–137. [CrossRef]
53. Tai, Z.; Yan, X.; Lang, J.; Xue, Q. Enhancement of capacitance performance of flexible carbon nanofiber paper by adding graphene nanosheets. *J. Power Sources* **2012**, *199*, 373–378. [CrossRef]
54. Vadukumpully, S.; Paul, J.; Mahanta, N.; Valiyaveettil, S. Flexible conductive graphene/poly(vinyl chloride) composite thin films with high mechanical strength and thermal stability. *Carbon* **2011**, *49*, 198–205. [CrossRef]
55. Wang, D.-W.; Li, F.; Zhao, J.; Ren, W.; Chen, Z.-G.; Tan, J.; Wu, Z.-S.; Gentle, I.; Lu, G.Q.; Cheng, H.-M. Fabrication of graphene/polyaniline composite paper via in situ anodic electropolymerization for high-performance flexible electrode. *ACS Nano* **2009**, *3*, 1745–1752. [CrossRef]
56. Wang, R.; Bian, H.; Ji, H.; Yang, R. Preparation of lignocellulose/graphene composite conductive paper. *Cellulose* **2018**, *25*, 6139–6149. [CrossRef]
57. Pan, L.; Xia, Y.; Qiu, B.; Zhao, H.; Guo, H.; Jia, K.; Gu, Q.; Liu, Z. Synthesis and electrochemical performance of micro-sized Li-rich layered cathode material for Lithium-ion batteries. *Electrochim. Acta* **2016**, *211*, 507–514. [CrossRef]
58. Xu, Y.; Hong, W.; Bai, H.; Li, C.; Shi, G. Strong and ductile poly (vinyl alcohol)/graphene oxide composite films with a layered structure. *Carbon* **2009**, *47*, 3538–3543. [CrossRef]
59. Xu, J.; Wang, K.; Zu, S.-Z.; Han, B.-H.; Wei, Z. Hierarchical nanocomposites of polyaniline nanowire arrays on graphene oxide sheets with synergistic effect for energy storage. *ACS Nano* **2010**, *4*, 5019–5026. [CrossRef]
60. Wang, L.; Ye, Y.; Lu, X.; Wen, Z.; Li, Z.; Hou, H.; Song, Y. Hierarchical nanocomposites of polyaniline nanowire arrays on reduced graphene oxide sheets for supercapacitors. *Sci. Rep.* **2013**, *3*, 3568. [CrossRef]
61. Xu, C.; Cui, A.; Xu, Y.; Fu, X. Graphene oxide–TiO$_2$ composite filtration membranes and their potential application for water purification. *Carbon* **2013**, *62*, 465–471. [CrossRef]
62. Yu, C.; Ma, P.; Zhou, X.; Wang, A.; Qian, T.; Wu, S.; Chen, Q. All-solid-state flexible supercapacitors based on highly dispersed polypyrrole nanowire and reduced graphene oxide composites. *ACS Appl. Mater. Interfaces* **2014**, *6*, 17937–17943. [CrossRef]
63. Yun, Y.J.; Hong, W.G.; Kim, W.J.; Jun, Y.; Kim, B.H. A novel method for applying reduced graphene oxide directly to electronic textiles from yarns to fabrics. *Adv. Mater.* **2013**, *25*, 5701–5705. [CrossRef]
64. Zhang, X.-J.; Wang, G.-S.; Cao, W.-Q.; Wei, Y.-Z.; Liang, J.-F.; Guo, L.; Cao, M.-S. Enhanced microwave absorption property of reduced graphene oxide (RGO)-MnFe2O4 nanocomposites and polyvinylidene fluoride. *ACS Appl. Mater. Interfaces* **2014**, *6*, 7471–7478. [CrossRef]
65. Zhou, X.; Chen, Q.; Wang, A.; Xu, J.; Wu, S.; Shen, J. Bamboo-like composites of V$_2$O$_5$/polyindole and activated carbon cloth as electrodes for all-solid-state flexible asymmetric supercapacitors. *ACS Appl. Mater. Interfaces* **2016**, *8*, 3776–3783. [CrossRef] [PubMed]
66. Xu, Y.; Wang, L.; Cao, P.; Cai, C.; Fu, Y.; Ma, X. Mesoporous composite nickel cobalt oxide/graphene oxide synthesized via a template-assistant co-precipitation route as electrode material for supercapacitors. *J. Power Sources* **2016**, *306*, 742–752. [CrossRef]

67. Zhao, C.; Wang, Q.; Zhang, H.; Passerini, S.; Qian, X. Two-dimensional titanium carbide/RGO composite for high-performance supercapacitors. *ACS Appl. Mater. Interfaces* **2016**, *8*, 15661–15667. [CrossRef] [PubMed]
68. Han, Z.J.; Seo, D.H.; Yick, S.; Chen, J.H.; Ostrikov, K.K. MnO$_x$/carbon nanotube/reduced graphene oxide nanohybrids as high-performance supercapacitor electrodes. *NPG Asia Mater.* **2014**, *6*, e140. [CrossRef]
69. Cao, X.; Zheng, B.; Shi, W.; Yang, J.; Fan, Z.; Luo, Z.; Rui, X.; Chen, B.; Yan, Q.; Zhang, H. Reduced graphene oxide-wrapped MoO$_3$ composites prepared by using metal–organic frameworks as precursor for all-solid-state flexible supercapacitors. *Adv. Mater.* **2015**, *27*, 4695–4701. [CrossRef]
70. Dong, X.; Wang, K.; Zhao, C.; Qian, X.; Chen, S.; Li, Z.; Liu, H.; Dou, S. Direct synthesis of RGO/Cu$_2$O composite films on Cu foil for supercapacitors. *J. Alloys Compd.* **2014**, *586*, 745–753. [CrossRef]
71. Ghorbani, M.; Golobostanfard, M.R.; Abdizadeh, H. Flexible freestanding sandwich type ZnO/rGO/ZnO electrode for wearable supercapacitor. *Appl. Surf. Sci.* **2017**, *419*, 277–285. [CrossRef]
72. Hu, Y.; Guan, C.; Ke, Q.; Yow, Z.F.; Cheng, C.; Wang, J. Hybrid Fe$_2$O$_3$ nanoparticle clusters/rGO paper as an effective negative electrode for flexible supercapacitors. *Chem. Mater.* **2016**, *28*, 7296–7303. [CrossRef]
73. Ali, G.A.M.; Yusoff, M.M.; Algarni, H.; Chong, K.F. One-step electrosynthesis of MnO$_2$/rGO nanocomposite and its enhanced electrochemical performance. *Ceram. Int.* **2018**, *44*, 7799–7807. [CrossRef]
74. Thalji, M.R.; Ali, G.A.M.; Liu, P.; Zhong, Y.L.; Chong, K.F. W$_{18}$O$_{49}$ nanowires-graphene nanocomposite for asymmetric supercapacitors employing AlCl$_3$ aqueous electrolyte. *Chem. Eng. J.* **2021**, *409*, 128216. [CrossRef]
75. Kabiri, R.; Namazi, H. Nanocrystalline cellulose acetate (NCCA)/graphene oxide (GO) nanocomposites with enhanced mechanical properties and barrier against water vapor. *Cellulose* **2014**, *21*, 3527–3539. [CrossRef]
76. Kafy, A.; Sadasivuni, K.K.; Kim, H.-C.; Akther, A.; Kim, J. Designing flexible energy and memory storage materials using cellulose modified graphene oxide nanocomposites. *Phys. Chem. Chem. Phys.* **2015**, *17*, 5923–5931. [CrossRef]
77. Kim, H.; Cho, M.Y.; Kim, M.H.; Park, K.Y.; Gwon, H.; Lee, Y.; Roh, K.C.; Kang, K. A novel high-energy hybrid supercapacitor with an anatase TiO$_2$-reduced graphene oxide anode and an activated carbon cathode. *Adv. Energy Mater.* **2013**, *3*, 1500–1506. [CrossRef]
78. Mianehrow, H.; Sabury, S.; Bazargan, A.; Sharif, F.; Mazinani, S. A flexible electrode based on recycled paper pulp and reduced graphene oxide composite. *J. Mater. Sci. Mater. Electron.* **2017**, *28*, 4990–4996. [CrossRef]

Article

A Facile Synthesis of Noble-Metal-Free Catalyst Based on Nitrogen Doped Graphene Oxide for Oxygen Reduction Reaction

Vladimir P. Vasiliev [1,*], Roman A. Manzhos [1], Valeriy K. Kochergin [1], Alexander G. Krivenko [1], Eugene N. Kabachkov [1,2], Alexander V. Kulikov [1], Yury M. Shulga [1] and Gennady L. Gutsev [3,*]

[1] Institute of Problems of Chemical Physics of RAS, Acad. Semenov ave. 1, 142432 Chernogolovka, Russia; rmanzhos@yandex.ru (R.A.M.); kocherginvk@yandex.ru (V.K.K.); krivenko@icp.ac.ru (A.G.K.); en.kabachkov@gmail.com (E.N.K.); kulav@icp.ac.ru (A.V.K.); yshulga@gmail.com (Y.M.S.)
[2] Chernogolovka Scientific Center, Russian Academy of Sciences, 142432 Chernogolovka, Russia
[3] Department of Physics, Florida A&M University, Tallahassee, FL 32307, USA
* Correspondence: vpvasiliev@mail.ru (V.P.V.); gennady.gutsev@famu.edu (G.L.G.)

Abstract: A simple method for the mechanochemical synthesis of an effective metal-free electrocatalyst for the oxygen reduction reaction was demonstrated. A nitrogen-doped carbon material was obtained by grinding a mixture of graphene oxide and melamine in a planetary ball mill. The resulting material was characterized by XPS, EPR, and Raman and IR spectroscopy. The nitrogen concentration on the N-bmGO surface was 5.5 at.%. The nitrogen-enriched graphene material (NbmGO has half-wave potential of −0.175/−0.09 V and was shown to possess high activity as an electrocatalyst for oxygen reduction reaction. The electrocatalytic activity of NbmGO can be associated with a high concentration of active sites for the adsorption of oxygen molecules on its surface. The high current retention (93% for 12 h) after continuous polarization demonstrates the excellent long-term stability of NbmGO.

Keywords: oxygen reduction reaction; noble-metal-free catalysts; graphene oxide; melamine; ball-milling; N-doped

1. Introduction

Fuel cells are considered as promising, renewable, and environmentally friendly energy sources. For the widespread use of fuel cells, however, it is necessary to solve several problems, one of which is related with fabrication of proper catalysts. Currently, in fuel cells with polymer electrolyte membranes, the active component of cathodic catalysts is presented by nanoparticles composed of platinum or its alloys which are deposited on carbon black [1–3]. Catalysts of this type have both advantages such as complete oxygen reduction and low overvoltage and obvious and fundamentally inevitable disadvantages such as sensitivity to impurities, high cost, limited resources, and low oxygen reduction reaction rates (ORR). It should be noted that the latter property restricts the load characteristics of fuel cells based on such catalysts [4–6]. One of the directions of research in the creation of ORR catalysts is the use of various types of carbon nanoforms (graphene-like structures, nanotubes, fullerenes, etc.) as carriers for Pt and its alloys. Another direction of research is based on the development of simple and effective methods for modifying the carbon structures themselves to create on their basis metal-free catalysts for the oxygen electroreduction in fuel cells.

Carbon nanostructures doped with p-elements are considered to be promising electrocatalysts for the oxygen reduction reaction [1,7,8]. The catalytic characteristics of modified graphene-like structures, for example, with nitrogen, are due to ability of their atoms to form a delocalized conjugated system with sp^2-hybridized carbon, where a common positive charge on atoms of the carbon frame adjacent to nitrogen is created [9,10]. Note that in order to increase the productivity of catalysts based on graphene structures, it is necessary to increase the specific concentration of active ORR centers on the electrode surface accessible to electrolyte.

It was shown in many studies (see, for example, [8,11,12]), that the oxygen reduction reaction proceeds on such structures according to the four-electron mechanism, which was previously considered [13,14] to be characteristic only for platinum and platinum alloy catalysts. The production of nitrogen-doped carbon nanoform using standard methods is complicated by the possible toxicity of nitrogen precursors and their contamination of the final products as well as by the necessity to use expensive specialized equipment is required. Therefore, the development of non-standard methods for obtaining doped graphene-based catalysts presents an important task. In particular, a mechanochemical approach can be used since it allows doping of graphene structures with nitrogen atoms to be performed. In our case, it allows to create active ORR centers distributed over highly dispersed material. In other words, there is an increased electrode surface available for electrolyte. It should be noted that particles of such materials are less susceptible to agglomeration, in contrast to hydrophobic graphene sheets, which should contribute to a higher temporal stability of the electrocatalytic characteristics of such catalysts.

In the present work, we proposed a strategy for a simple one-step solid-state synthesis of nitrogen-enriched carbon powder (NbmGO) using inexpensive industrial precursors, namely graphene oxide (GO) and melamine. The synthesized material was characterized by methods of scanning electron microscopy, X-ray photoelectron spectroscopy, and infrared and Raman spectroscopies. In addition, NbmGO was tested as an ORR electrocatalyst and showed a higher efficiency (a decrease in ORR overvoltage and an increase in the contribution of complete oxygen reduction to water in the overall process) compared to the results obtained previously [15]. It should be noted that the number of studies on the synthesis of carbon materials and their modification using solid-phase methods is quite scarce, and restricted to the splitting of graphite [16,17] and the preparation of composites of carbon materials with transition metal oxides [18,19].

2. Experimental

2.1. Synthesis of Nitrogen-Doped Carbon Material

Graphene oxide was synthesized using a modified Hammers method [20] with chemical composition $C_8O_{4.6}H_{1.8}(H_2O)_{0.58}$ and density ~1.2 g/cm^3. Melamine $C_3N_6H_6$ (99.9%, BASF SE, *Mannheim*, Germany) was used as a source of nitrogen (for details, see the Supplementary Materials). The mechanochemical synthesis was carried out in a FRITSCH pulverisette-6 planetary mill with a grinding vessel and balls made of ZrO_2. The internal diameter of the grinding vessel, the volume, and the ball diameter were 65 mm, 85 mL, and 5 mm, respectively. The GO/melamine ratio of reagents was 4:1, rotation speed was 400 rpm, and grinding time was 6 min. After grinding, the resulting powder was kept for 1 h in a 10% aqueous solution of ammonia, treated in an ultrasonic bath, and then it was centrifuged and washed with water 4–5 times to remove melamine residues.

2.2. Characterization

The sample images were acquired using a Zeiss LEO SUPRA-25 scanning electron microscope (Jena, Germany), and Raman spectra were recorded using a Bruker Senterra spectrometer Billerica, Billerica, MA, USA). The laser radiation wavelength was 532 nm, the radiation power at the measurement point was 1 mW, and the diameter of the analyzing laser beam was ~1 μm. Infrared spectra of the powders were obtained using an FT-IR VERTEX 70v spectrometer (Billerica, MA, USA) in vacuum (50 scans with the resolution of 4 cm^{-1}).

XPS spectra were obtained with the use of an electronic spectrometer for chemical analysis Specs PHOIBOS 150 MCD (Berlin, Germany). When recording the spectra, the vacuum in the spectrometer chamber did not exceed 2×10^{-10} Torr; the X-ray tube was equipped with a magnesium anode (Mg Kα radiation is 1253.6 eV) and the source power was 225 W. The survey spectrum was recorded in the range 0–1000 eV in the constant transmission energy mode (40 eV for the survey spectrum and 10 eV for individual lines). The survey spectrum was recorded with a step of 1 eV, while the spectra with individual lines with a step of 0.05 eV.

The ESR spectra of the powders were recorded at room temperature with a Bruker Elexsys II E 500 EPR spectrometer (Billerica, MA, USA) and an SE/X 2544 radio spectrometer (Radiopan, Poznan, Poland). The number of spins N and the g-factor were determined using the Xepr software package. To check the correctness of these procedures, a weighed quantity of $CuSO_4 \times 5H_2O$ and a DPPH sample with a g-factor of 2.0036 were used. The accuracy of concentration determination was ~15%. The electronic absorption spectra were obtained using a spectrophotometer PE-5400uf (Orenburg, Russia) and a Shimadzu UV-3101PC. The conductivity of the sample films was recorded on a potentiostat P-20X Elins (Orenburg, Russia) using a Micru XIDE1 thin-film Au-interdigitated electrode (90 pairs, 10/10 µm, electrode/gap).

2.3. Electrochemical Measurements

Voltammograms with linear potential sweep were measured in a three-electrode cell on a setup with the RRDE-3A rotating disk electrode s (ALS Co., Ltd., Naka-ku Sakai, Japan) using a potentiostat Autolab PGSTAT 302N (Metrohm Autolab, Utrecht, Holland) in an oxygen-saturated 0.1 M KOH solution with a potential sweep rate of v = 10 mV/s at electrode rotation frequencies ω = 360–6400 rpm. The current–voltage curves were analyzed using the Koutetsky–Levich equation [21]:

$$\frac{1}{j} = \frac{1}{j_k} + \frac{1}{j_d} \tag{1}$$

$$j_k = nFkc^0 \tag{2}$$

$$j_d = 0.62nFD^{2/3}\omega^{1/2}v^{-1/6}c^0 \tag{3}$$

where j_k is the density of kinetic current, j_d is the density of limiting diffusion current, F is the Faraday number (F = 96,485 C/mol), n is the number of electrons participating in the electrode reaction, D is the coefficient of oxygen diffusion in a 0.1 M KOH solution (D = 1.9×10^{-5} cm^2/s), v is the kinematic viscosity of a 0.1 M KOH solution (v = 0.01 cm^2/s), and c^0 is the volume concentration of dissolved oxygen (c^0 = 1.2 mM in a 0.1 M KOH solution) [9,22].

A glassy carbon (GC) disk with a diameter of 3 mm, pressed into a PEEK polymer (ALS Co., Ltd., Naka-ku Sakai, Japan), served as a working electrode. The electrode surface was preliminarily polished with 1µm Al_2O_3 powder, then a drop of an aqueous suspension of bmGO or NbmGO with a volume of ~ 6 µL and a concentration of 1 mg/mL, containing 0.01 wt% Nafion, was applied and dried at room temperature. A platinum wire with an area of ~1 cm^2 was used as an auxiliary electrode and an Ag/AgCl electrode filled with a saturated KCl solution was used as a reference electrode. All potentials (E) are given on the scale of the reference electrode. The bmGO deposited on the GC electrode was preliminarily electrochemically reduced during potential cycling (20–50 cycles) in an O_2-saturated 0.1 M KOH solution in the E range from 50 mV to $-$1300 mV at a potential sweep rate of 50 mV/s. The catalyst stability was tested using chronoamperometry method at $-$250 mV for 12 h.

3. Results and Discussions

3.1. Structural Characterization of bmGO and NbmGO

3.1.1. SEM

Figure 1 shows SEM images of samples of the starting graphene oxide (GO), ball milled graphene oxide (bmGO), and nitrogen-enriched carbon material (NbmGO) obtained by processing a mixture of GO and melamine in a ball mill. As can be seen from the figure, highly dispersed materials with a size of visible aggregates not exceeding 50 nm were formed by processing in the ball mill. This process increases the effective surface of the electrode accessible for electrolyte, and only small fragments of the original GO sheets can be observed in Figure 1b,c.

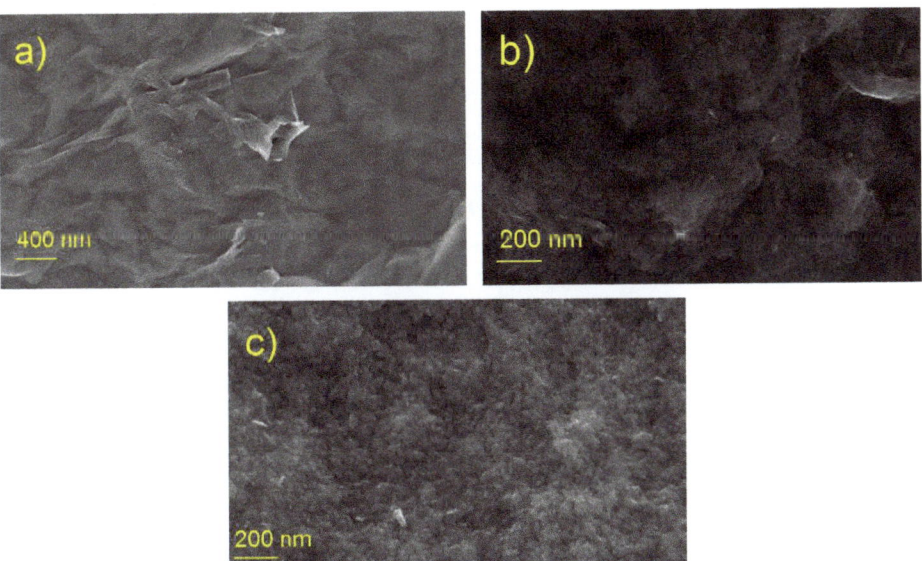

Figure 1. SEM images of GO (**a**), bmGO (**b**), and NbmGO (**c**).

The elemental composition of carbon materials was calculated using analytical lines of the survey XPS spectrum (Figure S1 of the Supplementary Materials). The oxygen concentration decreases markedly only for the NbmGO sample. The recovery of the sample is accompanied by an increase in the nitrogen content up to 5.5 at.% (Table S1 of the Supplementary Materials). The relative concentration of carbon in the NbmGO and bmGO samples increases insignificantly, by only 2.4–2.5 at.%. Although parent GO is an insulator, significant electronic conduction occurs in the films of the bmGO and NbmGO samples (Figure S2 of the Supplementary Materials).

3.1.2. XPS

The shape of the C1s line in the spectra of the samples under study and the line fitting with four symmetrical Gaussian–Lorentz curves are presented in Figure 2a–c. The position of the main most intense peak (C1) is typical for sp^2-carbon materials (Table S2 of the Supplementary Materials). The second most intense peak (C2) refers to carbon atoms that singly bonded with a nitrogen or oxygen atom of a hydroxyl (C–OH) or epoxy groups. The third peak (C3) can be attributed to the carbon atoms having two bonds with the oxygen atom (C=O or O–C–O). The last peak (C4) was assigned to the carbon atoms of the carboxyl groups (O–C=O) [23]. Note that the assignment of individual peaks was done in accordance with recommendations in [24].

According to the previous studies [8,25], nitrogen in a graphite-like matrix can be found in four or five configurations: pyridinic (N1, six-membered ring), pyrrolic (N2, five-membered ring), graphitic (N3/N4), and oxidized pyridinic (N5) (see Figure S3 in the Supplementary Materials). The pyridinic and pyrrolic nitrogen atoms are located at the edges of a graphene sheet or at the defect sites. Nitrogen atoms of the N3 and N4 types replace carbon atoms in the graphite structure and differ by the location type. The nitrogen atom scan be located at the edges of a graphene sheet (N3) or in its center (N4). Nitrogen, which is a part of cyano and amino groups, may also be present [26].

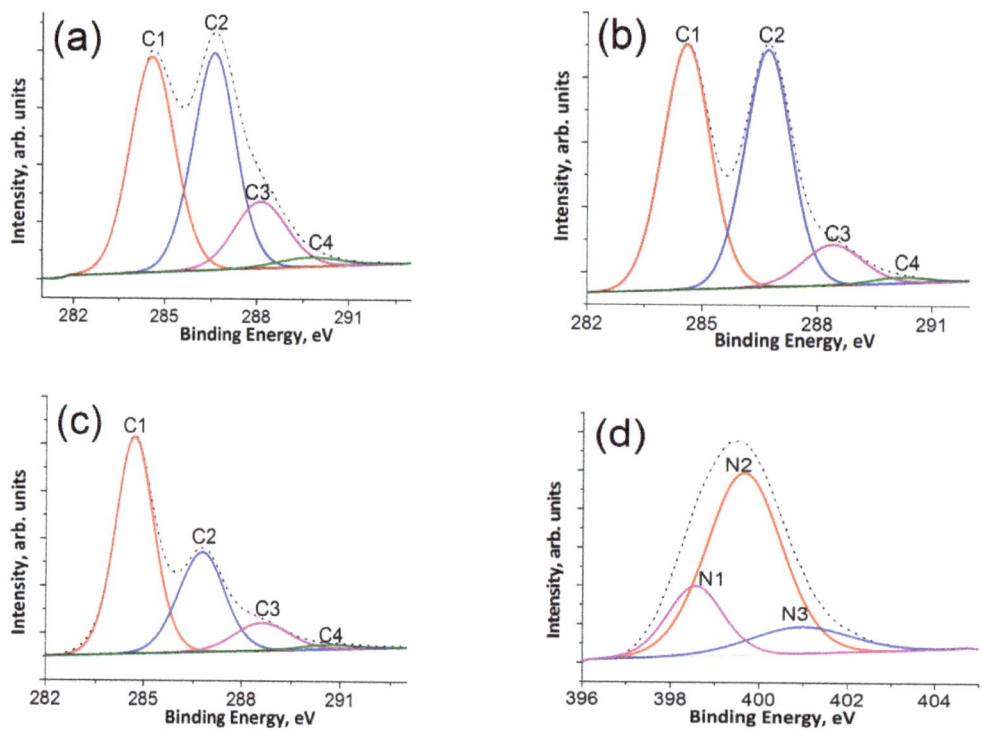

Figure 2. C1s lines in the XPS spectra of GO (**a**), bmGO (**b**), NbmGO (**c**), and N 1s lines of NbmGO (**d**).

The identification of surface nitrogen-containing groups can be carried out on the basis of an analysis of the fine structure of the N1s line in the high resolution XPS spectrum. According to the literature (see, e.g., [8,25] and references therein), pyridine nitrogen (N1) appears in the range 398.0–399.3 eV and pyrrole nitrogen (N2) appear in the range 399.8–401.2 eV) in the XPS spectra of nitrogen-doped carbon materials. It is worth noting that the N1s lines of amino (399.1 eV [25]) and cyano (399.3 eV [25]) groups are also located in this region; therefore, it is difficult to clearly identify it. The peak corresponding to the nitrogen atoms of the N4 type (inside a graphite sheet) is located at about 401 eV, and the peak corresponding to terminal graphite nitrogen (N3) is at 402.3 eV. Oxidized pyridine nitrogen (N5) corresponds to a peak at 402.8 eV while pyrrole nitrogen (N2) corresponds to a peak at 404.7 eV [25].

Three peaks can be distinguished in the N1s line of the NbmGO sample (see Figure 2d). When assigning these peaks, we came to a quite unexpected conclusion that the major contribution to the N1s line comes from the nitrogen atoms of the pyrrolic (N2) nitrogen. It is interesting to note that it was concluded in a model catalyst study that pyridinic nitrogen in graphite structures creates active centers for the oxygen reduction reaction [27].

3.1.3. FTIR

The FTIR spectra of melamine, initial graphene oxide, bmGO, and NbmGO are shown in Figure 3. It can be seen in the Figure that the IR spectrum of NbmGO differs from the IR spectra of the starting materials. Thus, absorption bands of stretching vibrations of N–H bonds, whose maxima in the spectrum of pure melamine are located at 3468, 3417, 3324, and 3121 cm^{-1}, are lacking in the NbmGO spectrum. At the same time, the spectrum of NbmGO has absorption bands, which can be attributed to stretching vibrations of cyano groups in different environments (the region from 2350 to 1900 cm^{-1}). Comparing the spectrum of NbmGO with the spectra of GO and bmGO, one can note that the absorption band due to the stretching C=O vibrations is practically absent in the NbmGO spectrum. Furthermore, one can notice a significant increase in the intensity of the absorption band associated with the vibrations of the C=C double bonds forbidden in the IR spectrum. According to the literature data, the band at 1360–1370 cm^{-1} can be attributed to vibrations of the C–OH bond, and the band at 1220 cm^{-1} corresponds to vibrations of C–O–C bonds. The band at 1060 cm^{-1} is attributed to the vibrations of alkoxy groups [28] (See Table S5 in the Supplementary Materials).

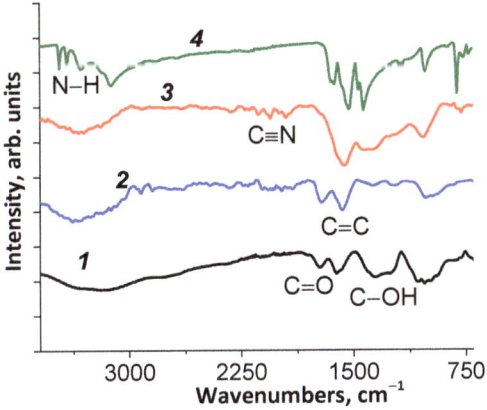

Figure 3. FTIR spectra of GO (*1*), bmGO (*2*), NbmGO (*3*), and melamine (*4*).

3.1.4. Raman

The Raman spectra of the samples studied are shown in Figure 4. The spectra contain peaks designated as D, G, and 2D. It is well known that the 2D peak for single-layer graphene is a narrow peak, whose intensity exceeds the intensity of peak G. In the case of two-layer graphene, the intensity of the 2D peak decreases, while its half-width increases [29]. When the number of layers in graphene becomes more than five, the 2D peak disappears.

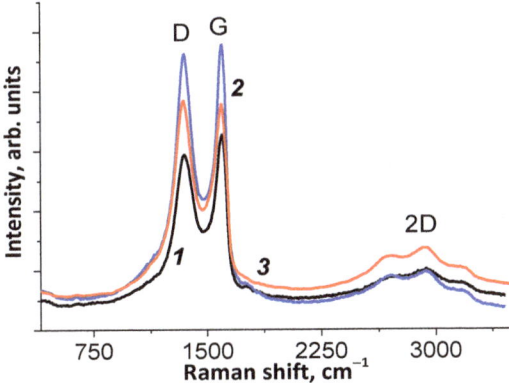

Figure 4. Raman spectra of GO (*1*), bmGO (*2*), and NbmGO (*3*). See the text for the peak designation.

The presence of a 2D peak in the Raman spectra of our samples means that there are few-layer graphene-like structures. In addition, note the shift in the position of the G peak towards lower values in the GO→bmGO→NbmGO series (see Table S3 of the

Supplementary Materials). A shift in the position of the G peak towards lower values during GO milling was also observed previously [30].

The ratio of the D and G band intensities (I_D/I_G) can be used to estimate the size of the sp^2 domains of L_a in the basal plane [31]:

$$L_a = (2.4 \times 10^{-10}) \lambda^4 (I_D/I_G)^{-1} \qquad (4)$$

Based on the ratio of the D/G peak intensities (see Table S3 of the Supplementary Materials), the size of sp^2 domains is 19 nm (NbmGO), 20 nm (bmGO), and 22 nm (GO). Thus, milling leads to a slight increase in the number of defects and a decrease in the size of sp^2 domains. The insertion of nitrogen atoms into the graphene oxide lattice also increases the number of defects and decreases the size of the graphene sp^2 domains.

3.1.5. ESR

The ESR spectra obtained for the samples under study are presented in Figure 5. As can be seen in the figure, the ESR spectra of all samples contains narrow singlet lines. The g factors of all samples are close to 2.0, which is typical for radicals where unpaired electrons of aromatic rings composed of carbon atoms occupy localized π-states [32,33].

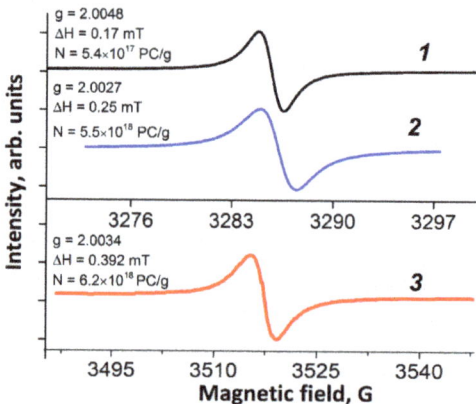

Figure 5. ESR spectra of GO (*1*), bmGO (*2*), and NbmGO (*3*) obtained at room temperature.

The measured spin concentration (see Table S4 of the Supplementary Materials) for the NbmGO sample is 6.2×10^{18} PC/g, which is an order of magnitude higher than the concentration of the GO sample (5.4×10^{17} PC/g) and is approximately comparable to the spin concentration for the bmGO sample (5.5×10^{18} PC/g). This is an indication of a significantly larger number of defects on the surface of carbon structures of the samples processed in the planetary mill, which largely determines their electrocatalytic activity in ORR.

3.2. Electrochemical Analysis

The voltammogram dependencies obtained on the initial GC electrode and GC electrodes coated with bmGO and NbmGO in a 0.1 M KOH solution saturated with oxygen are shown in Figure 6a. To determine the number of electrons n participating in the ORR (see Figure 6b), the j and E dependences were measured at different speeds of electrode rotation. Figure 6c shows a series of such voltammogram curves for NbmGO and Figure 6d displays the dependence of j on ω plotted in the Koutetsky–Levich coordinates. From the slope of the j-ω dependences, the values of n were calculated at various values of E (see Figure 6b).

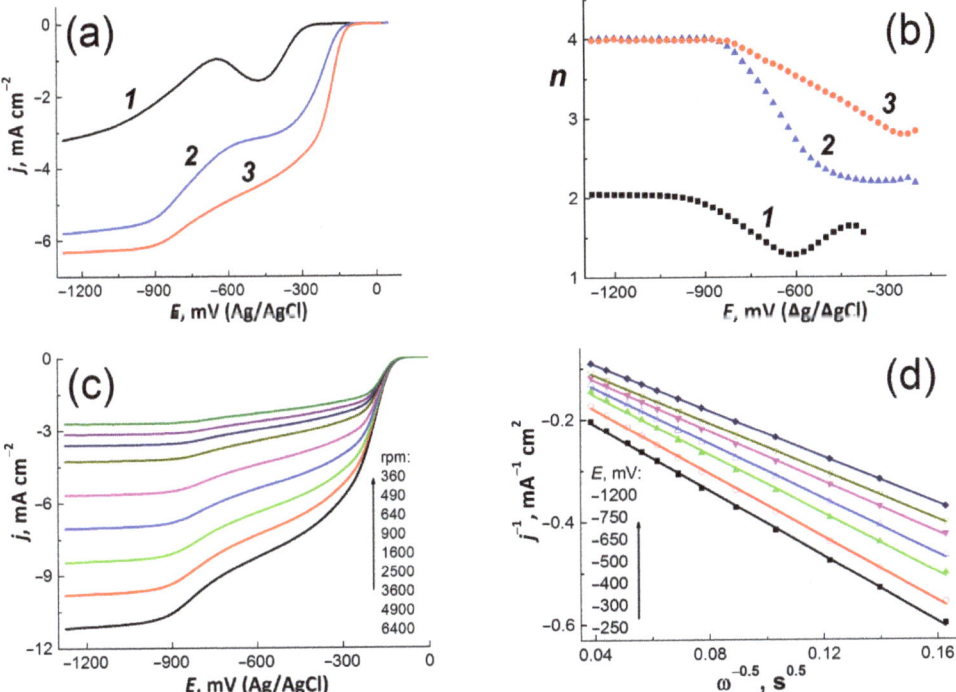

Figure 6. (a) Linear sweep voltammograms of O_2-saturated 0.1 M KOH solution for the bare GC electrode (curve 1) and GC coated with bmGO (curve 2) and NbmGO (curve 3), v = 10 mV/s, ω = 2000 rpm. (b) Dependencies of the electron transfer number n on the potential E for the bare GC electrode (curve 1), GC covered with bmGO (curve 2), and NbmGO (curve 3). (c) Voltammograms for NbmGO measured at different speed of electrode rotations. (d) Corresponding j-ω dependences in the Koutecký–Levich coordinates.

As can been seen from Figure 6a, the overpotential of the oxygen reduction reaction for bmGO and NbmGO significantly decreases compared to that of the initial GC. The half-wave potentials of the first oxygen reduction wave for GC, bmGO, and NbmGO are −365 mV, −225 mV, and −175 mV, respectively. In addition, an increase in the oxygen reduction current was observed when passing from the original GC electrode to the GC coated with bmGO and NbmGO.

On the voltammogram curve for bmGO (curve 2 in Figure 6a), two distinct waves which correspond to the predominant reduction of oxygen to hydrogen peroxide in the potential range from −250 to −500 mV ($n \approx$ 2.2–2.3) and water at E < −850 mV ($n \approx$ 4) can be distinguished. In the case of NbmGO (curve 3 in Figure 6a), the first wave includes a segment of linear growth of the cathodic current of in the potential range from −400 to −750 mV, while the second wave is characterized by reaching the limit at E < −900 mV and corresponds to the limiting diffusion current of complete oxygen reduction to water. The corresponding value of the limiting diffusion current density calculated by using Equation (3) is j_d = −6.4 m /cm^2. It should be noted that there is also a linear increase in the number of electrons participating in the oxygen reduction reaction from 2.8 to 4.0 (Figure 6b) in the potential range from −250 to −850 mV.

In the case of NbmGO O_2 is reduced to H_2O ($n \approx$ 2.8) even at low overpotentials (E = −200 mV) in addition to its reduction to hydrogen peroxide with a gradual increase in the contribution of this process until the complete reduction of oxygen to water at

$E < -850$ mV ($n \approx 4$). In general, bmGO and NbmGO are characterized by a significant decrease in the ORR overvoltage and higher n values as compared to glassy carbon. This is an indicator of a high concentration of active centers for the adsorption of both oxygen molecules on the surface of bmGO and NbmGO and intermediates of its reduction, which can be surface defects and edge regions of graphene-like structures [34], quinone groups [35], and pyridine nitrogen atoms in the case of the NbmGO sample [27,36], as shown earlier [37].

In addition to high catalytic performance, the ideal electrode material should have excellent long-term stability, which can be evaluated by prolonged chronoamperometry. A chronoamperometry test for the NbmGO catalyst was carried out at 2000 rpm in O_2-saturated 0.1 M KOH solution (see Figure S5 in the Supplementary Materials). The high current retention of 93 % after continuous polarization at -250 mV during 720 min clearly demonstrates the excellent stability of NbmGO.

4. Conclusions

The present work reports on a simple solid-phase method for the synthesis of an effective metal-free electrocatalyst for the oxygen reduction reaction from inexpensive industrial materials, namely, graphene oxide, and melamine. Our method allows the doping of graphene structures with nitrogen atoms and obtaining highly dispersed materials with increased effective electrode surfaces accessible by electrolytes. For the material obtained, a decrease in the ORR overvoltage and an increase in the contribution of the complete reduction of oxygen to water compared to those for glassy carbon are shown. The observed catalytic activity is determined by the high concentration of active sites for adsorption of oxygen molecules and intermediate intermediates of its reduction. The NbmGO long-term stability can be attributed to a small loss of active sites during the test.

Supplementary Materials: The following supporting information can be downloaded at: https://www.mdpi.com/article/10.3390/ma15030821/s1, Figure S1: XPS survey spectra of the GO, bmGO, and NbmGO samples, Table S1. Elemental composition (in at.%) of the samples under study; Figure S2: The conductivity of GO, bmGO, and NbmGO films at different $RH\%$; Figure S3: A model structure of NbmGO; Figure S4: Electronic absorption spectra of GO, bmGO and NbmGO films (quartz glass). The maximum of the electronic absorption spectrum shifts to shorter wavelengths (224 nm → 205 nm) for the samples of GO, bmGO and NbmGO and there is an increase in absorption observed in the range 400–800 nm, which is typical for sp^2-carbon; Table S2: Positions (E_b), full widths at half maximum ($FWHM$), and intensities (Int) of the peaks in the XPS spectra of the GO, bmGO and NbmGO samples; Table S3: The characteristic features in the FTIR spectra of GO and melamine; Table S4: Peak positions and the band intensity ratios (I_D/I_G) in the Raman spectra of graphite, GO, bmGO, and NbmGO; Table S5: ESR data obtained for samples of GO, bmGO, and NbmGO at room temperature; Table S6: Electrochemical properties of N-doped carbon materials; Figure S5: Long-term stability of NbmGO via chronoamperometry test at -300 mV in 0.1 M KOH and at room temperature, ω = 2000 rpm.

Author Contributions: Investigation, formal analysis, writing—original draft, project administration, V.P.V.; Investigation, writing, review and editing, R.A.M.; Investigation, V.K.K.; Writing, review, and editing, A.G.K.; Formal analysis, E.N.K.; Investigation, formal analysis, A.V.K.; Review and editing, project administration, Y.M.S.; Review and editing, data curation, G.L.G. All authors have read and agreed to the published version of the manuscript.

Funding: The study was performed in accordance with the State Assignments NosAAAA-A19-119032690060-9 and AAAA-A19-119061890019-5.

Institutional Review Board Statement: Not applicable.

Informed Consent Statement: Not applicable.

Data Availability Statement: Data available in a publicly accessible repository.

Acknowledgments: The work used the equipment of the Multi-User Analytical Center of IPCP RAS, and the Chernogolovka Scientific Center RAS.

Conflicts of Interest: The authors declare no conflict of interest.

References

1. Shao, M.H.; Chang, Q.W.; Dodelet, J.-P.; Chenitz, R. Recent advances in electrocatalysts for oxygen reduction reaction. *Chem. Rev.* **2016**, *116*, 3594–3657. [CrossRef] [PubMed]
2. Majlan, E.H.; Rohendi, D.; Daud, W.R.W.; Husaini, T.; Haque, M.A. Electrode for proton exchange membrane fuel cells: A review. *Renew. Sustain. Energy Rev.* **2018**, *89*, 117–134. [CrossRef]
3. Shao, Q.; Li, F.M.; Chen, Y.; Huang, X.Q. The advanced designs of high-performance platinum-based electrocatalysts: Recent progresses and challenges. *Adv. Mater. Interfaces* **2018**, *5*, 1800486. [CrossRef]
4. Gasteiger, H.A.; Kocha, S.S.; Sompalli, B.; Wagner, F.T. Activity benchmarks and requirements for Pt, Pt-alloy, and non-Pt oxygen reduction catalysts for PEMFCs. *Appl. Catal. B Environ.* **2005**, *56*, 9–35. [CrossRef]
5. Nørskov, J.K.; Rossmeisl, J.; Logadottir, A.; Lindqvist, L.; Kitchin, J.R.; Bligaard, T.; Jónsson, H. Origin of the overpotential for oxygen reduction at a fuel-cell cathode. *J. Phys. Chem. B* **2004**, *108*, 17886–17892. [CrossRef]
6. Greeley, J.; Stephens, I.E.L.; Bondarenko, A.S.; Johansson, T.P.; Hansen, H.A.; Jaramillo, T.F.; Rossmeisl, J.; Chorkendorff, I.; Norskov, J.K. Alloys of platinum and early transition metals as oxygen reduction electrocatalysts. *Nat. Chem.* **2009**, *1*, 552–556. [CrossRef] [PubMed]
7. Wang, Y.; Shao, Y.Y.; Matson, D.W.; Li, J.H.; Lin, Y.H. Nitrogen-doped graphene and its application in electrochemical biosensing. *ACS Nano* **2010**, *4*, 1790–1798. [CrossRef]
8. Daems, N.; Sheng, X.; Vankelecom, I.F.J.; Pescarmona, P.P. Metal-free doped carbon materials as electrocatalysts for the oxygen reduction reaction. *J. Mater. Chem. A* **2014**, *2*, 4085–4110. [CrossRef]
9. Qu, L.T.; Liu, Y.; Baek, J.B.; Dai, L.M. Nitrogen-doped graphene as efficient metal-free electrocatalyst for oxygen reduction in fuel cells. *ACS Nano* **2010**, *4*, 1321–1326. [CrossRef]
10. Gong, K.P.; Du, F.; Xia, Z.H.; Durstock, M.; Dai, L.M. Nitrogen doped carbon nanotube arrays with high electrocatalytic activity for oxygen reduction. *Science* **2009**, *323*, 760–764. [CrossRef]
11. Kakaei, K.; Ghadimi, G. A green method for nitrogen-doped graphene and its application for oxygen reduction reaction in alkaline media. *Mater. Technol.* **2020**, *36*, 46–53. [CrossRef]
12. Sheng, Z.-H.; Shao, L.; Chen, J.-J.; Bao, W.-J.; Wang, F.-B.; Xia, X.-H. Catalyst-free synthesis of nitrogen-doped graphene via thermal annealing graphite oxide with melamine and its excellent electrocatalysis. *ACS Nano* **2011**, *5*, 4350–4358. [CrossRef] [PubMed]
13. Jukk, K.; Kongi, N.; Tammeveski, K.; Arán-Ais, R.M.; Solla-Gullón, J.; Feliu, J.M. Loading effect of carbon-supported platinum nanocubes on oxygen electroreduction. *Electrochim. Acta* **2017**, *251*, 155–166. [CrossRef]
14. Gómez-Marín, A.M.; Feliu, J.M.; Ticianelli, E. Oxygen reduction on platinum surfaces in acid media: Experimental evidence of a CECE/DISP initial reaction path. *ACS Catal.* **2019**, *9*, 2238–2251. [CrossRef]
15. Manzhos, R.A.; Baskakov, S.A.; Kabachkov, E.N.; Korepanov, V.I.; Dremova, N.N.; Baskakova, Y.V.; Krivenko, A.G.; Shulga, Y.M.; Gutsev, G.L. Reduced graphene oxide aerogel inside melamine sponge as an electrocatalyst for the oxygen reduction reaction. *Materials* **2021**, *14*, 322. [CrossRef] [PubMed]
16. Leo, V.; Rodriguez, A.M.; Prieto, P.; Prato, M.; Vazquez, E. Exfoliation of graphite with triazine derivatives under ball-milling conditions: Preparation of few-layer graphene via selective noncovalent interactions. *ACS Nano* **2014**, *8*, 563–571. [CrossRef]
17. Venkatachalam, P.; Ganesan, S.; Rengapillai, S.; Vembu, S.; Sivakumar, M. Physicochemical exfoliation of graphene sheet using graphitic carbon nitride. *New J. Chem.* **2019**, *43*, 16200–16206. [CrossRef]
18. Ahmad, J.; Sofi, F.A.; Mehraj, O.; Majid, K. Fabrication of highly photocatalytic active anatase TiO$_2$-graphene oxide heterostructures via solid phase ball milling for environmental remediation. *Surf. Interfaces* **2018**, *13*, 186–195. [CrossRef]
19. Kahimbi, H.; Hong, S.B.; Yang, M.; Choi, B.G. Simultaneous synthesis of NiO/reduced graphene oxide composites by ball milling using bulk Ni and graphite oxide for supercapacitor applications. *J. Electroanal. Chem.* **2017**, *786*, 14–19.
20. Hummers, W.S.; Offeman, R.E. Preparation of Graphitic Oxide. *J. Am. Chem. Soc.* **1958**, *80*, 1339.
21. Bard, A.J.; Faulkner, L.R. *Electrochemical Methods: Fundamentals and Applications*, 2nd ed.; Wiley: New York, NY, USA, 2001.
22. Jürmann, G.; Tammeveski, K. Electroreduction of oxygen on multi-walled carbon nanotubes modified highly oriented pyrolytic graphite electrodes in alkaline solution. *J. Electroanal. Chem.* **2006**, *597*, 119–126. [CrossRef]
23. Hanifah, M.F.R.; Jaafar, J.; Aziz, M.; Ismail, A.F.; Othman, M.H.D.; Rahman, M.A.; Norddin, M.N.A.M.; Yusof, N.; Salleh, W.N.W. Efficient reduction of graphene oxide nanosheets using Na$_2$C$_2$O$_4$ as a reducing agent. *Funct. Mater. Lett.* **2015**, *8*, 155026. [CrossRef]
24. Lesiak, B.; Kövér, L.; Tóth, J.; Zemek, J.; Jiricek, P.; Kromka, A.; Rangam, N. C sp^2/sp^3 hybridisations in carbon nanomaterials—XPS and (X) AES study. *Appl. Surf. Sci.* **2018**, *452*, 223–231. [CrossRef]
25. Lazar, P.; Mach, R.; Otyepka, M. Spectroscopic fingerprints of graphitic, pyrrolic, pyridinic, and chemisorbed nitrogen in N-doped graphene. *J. Phys. Chem. C* **2019**, *123*, 10695–10702. [CrossRef]

26. Mondal, O.; Mitra, S.; Pal, M.; Datta, A.; Dhara, S.; Chakravorty, D. Reduced graphene oxide synthesis by high energy ball milling. *Mater. Chem. Phys.* **2015**, *161*, 123–129. [CrossRef]
27. Guo, D.; Shibuya, R.; Akiba, C.; Saji, S.; Kondo, T.; Nakamura, J. Active sites of nitrogen-doped carbon materials for oxygen reduction reaction clarified using model catalysts. *Science* **2016**, *351*, 361–365. [CrossRef]
28. Hanifah, M.F.R.; Jaafar, J.; Othman, M.H.D.; Ismail, A.F.; Rahman, M.A.; Yusof, N.; Salleh, W.N.W.; Aziz, F. Facile synthesis of highly favorable graphene oxide: Effect of oxidation degree on the structural, morphological, thermal and electrochemical properties. *Materialia* **2019**, *6*, 100344. [CrossRef]
29. Calizo, I.; Balandin, A.A.; Bao, W.; Miao, F.; Lau, C.N. Temperature dependence of the Raman spectra of graphene and graphene multilayers. *Nano Lett.* **2007**, *7*, 2645–2649. [CrossRef]
30. Fu, J.; Wei, C.; Wang, W.; Wei, J.L.; Lu, J. Studies of structure and properties of graphene oxide prepared by ball milling. *Mater. Res. Innov.* **2015**, *19*, S1-277–S1-280. [CrossRef]
31. Pimenta, M.A.; Dresselhaus, G.; Dresselhaus, M.S.; Cançado, L.G.; Jorio, A.; Saito, R. Studying disorder in graphite-based systems by Raman spectroscopy. *Phys. Chem. Chem. Phys.* **2007**, *9*, 1276–1290. [CrossRef]
32. Dvoranova, D.; Barbierikova, Z.; Mazur, M.; Garcia-Lopez, E.I.; Marci, G.; Luspai, K.; Brezova, V. EPR investigations of polymeric and H_2O_2-modified C_3N_4-based photocatalysts. *J. Photochem. Photobiol. A* **2019**, *375*, 100–113. [CrossRef]
33. Di, J.; Xia, J.X.; Li, X.W.; Ji, M.X.; Xu, H.; Chen, Z.G.; Li, H.M. Constructing confined surface carbon defects in ultrathin graphitic carbon nitride for photocatalytic free radical manipulation. *Carbon* **2016**, *107*, 1–10. [CrossRef]
34. Shen, A.L.; Zou, Y.Q.; Wang, Q.; Dryfe, R.A.W.; Huang, X.B.; Dou, S.; Dai, L.M.; Wang, S.Y. Oxygen reduction reaction in a droplet on graphite: Direct evidence that the edge is more active than the basal plane. *Angew. Chem. Int. Ed.* **2014**, *53*, 10804–10808. [CrossRef] [PubMed]
35. Vasiliev, V.P.; Kotkin, A.S.; Kochergin, V.K.; Manzhos, R.A.; Krivenko, A.G. Oxygen reduction reaction at few-layer graphene structures obtained via plasma-assisted electrochemical exfoliation of graphite. *J. Electroanal. Chem.* **2019**, *851*, 113440. [CrossRef]
36. Lai, L.F.; Potts, J.R.; Zhan, D.; Wang, L.; Poh, C.K.; Tang, C.H.; Gong, H.; Shen, Z.X.; Jianyi, L.Y.; Ruoff, R.S. Exploration of the active center structure of nitrogen-doped graphene-based catalysts for oxygen reduction reaction. *Energy Environ. Sci.* **2012**, *5*, 7936–7942. [CrossRef]
37. Vasiliev, V.P.; Manzhos, R.A.; Krivenko, A.G.; Kabachkov, E.N.; Shulga, Y.M. Nitrogen-enriched carbon powder prepared by ball-milling of graphene oxide with melamine: An efficient electrocatalyst for oxygen reduction reaction. *Mendeleev Commun.* **2021**, *31*, 529–531. [CrossRef]

MDPI
St. Alban-Anlage 66
4052 Basel
Switzerland
www.mdpi.com

Materials Editorial Office
E-mail: materials@mdpi.com
www.mdpi.com/journal/materials

Disclaimer/Publisher's Note: The statements, opinions and data contained in all publications are solely those of the individual author(s) and contributor(s) and not of MDPI and/or the editor(s). MDPI and/or the editor(s) disclaim responsibility for any injury to people or property resulting from any ideas, methods, instructions or products referred to in the content.

www.ingramcontent.com/pod-product-compliance
Lightning Source LLC
LaVergne TN
LVHW070639100526
838202LV00013B/840